# SpringerBriefs in Statistics

# SpringerBriefs in Statistics – ABE

SpringerBriefs present concise summaries of cutting-edge research and practical applications across a wide spectrum of fields. Featuring compact volumes of 50 to 125 pages, the series covers a range of content from professional to academic. Briefs are characterized by fast, global electronic dissemination, standard publishing contracts, standardized manuscript preparation and formatting guidelines, and expedited production schedules.

Typical topics might include:

- A timely report of state-of-art techniques
- A bridge between new research results, as published in journal articles, and a contextual literature review
- A snapshot of a hot or emerging topic
- An in-depth case study
- A presentation of core concepts that students must understand in order to make independent contributions

This SpringerBriefs in Statistics – ABE subseries aims to publish relevant contributions to several fields in the Statistical Sciences, produced by full members of the ABE – Associação Brasileira de Estatística (Brazilian Statistical Association), or by distinguished researchers from all Latin America. These texts are targeted to a broad audience that includes researchers, graduate students and professionals, and may cover a multitude of areas like Pure and Applied Statistics, Actuarial Science, Econometrics, Quality Assurance and Control, Computational Statistics, Risk and Probability, Educational Statistics, Data Mining, Big Data and Confidence and Survival Analysis, to name a few.

More about ABE: http://www.redeabe.org.br

Heleno Bolfarine • Mário de Castro • Manuel Galea

# Regression Models for the Comparison of Measurement Methods

ABE

Springer

Heleno Bolfarine
Department of Statistics
University of São Paulo
Sao Paulo, São Paulo, Brazil

Mário de Castro
Department of Applied Mathematics
and Statistics
University of São Paulo
São Carlos, São Paulo, Brazil

Manuel Galea
Department of Statistics
Pontificia Universidad Católica de Chile
Santiago, RM - Santiago, Chile

ISSN 2191-544X ISSN 2191-5458 (electronic)
SpringerBriefs in Statistics
ISSN 2524-6917 ISSN 2524-6925 (electronic)
SpringerBriefs in Statistics – ABE
ISBN 978-3-030-57934-0 ISBN 978-3-030-57935-7 (eBook)
https://doi.org/10.1007/978-3-030-57935-7

Mathematics Subject Classification: 62Jxx, 62Kxx, 62Pxx



This Springer imprint is published by the registered company Springer Nature Switzerland AG
The registered company address is: Gewerbestrasse 11, 6330 Cham, Switzerland

*To my wife Ana Maria Bolfarine*

(Heleno Bolfarine)

*To my mother* (in memoriam)

(Mário de Castro)

*To my parents Lorenza and Enrique* (in memoriam), *my wife Patricia, and my children Rodrigo and Felipe*

(Manuel Galea)

# Preface

In this book, we present models and statistical methods for the comparison of measurement methods problem. The need to compare measurement methods or systems, which differ in cost, speed, or other factors, has frequently appeared in various areas of knowledge, such as medicine, biology, analytical chemistry, geology, environmental sciences, psychology, and education, among many others.

In this book, we consider statistical models for continuous data, and unlike the current literature, we adopt functional measurement error models. This background includes the explicit formulation of the functional regression model and the proposal of statistical tests for detection of bias. For parameter estimation and hypothesis testing, we use the maximum likelihood method in the presence of incidental parameters and the corrected score methodology. To the best of our knowledge, both approaches have not been considered in the existing books on this topic. Furthermore, we include tools for the assessment of model adequacy and sensitivity analysis. Although these topics should be an essential part of the analytical method validation problem, they have received only scarce attention in the literature. As an essential element of the book, we furnish R code with functions implementing the techniques and examples, so that practitioners can easily analyze their own data sets.

This book is organized into five chapters. Chapter 1 provides a general introduction to the topics developed in this book. In Chap. 2, we present the statistical model and tools for statistical inference for the comparison of two measurement methods. The probability of agreement is also included. In Chap. 3, these techniques are extended to the general case comprising two or more measurement methods. Chapter 4 covers model checking and influence diagnostics methods. We embrace the local influence methodology to assess the sensitivity of the proposed test statistics to perturbations in the data set and/or in the assumptions underlying the statistical model. In Chap. 5, we illustrate the usefulness of the methodology presented in the previous chapters with real data sets. In Appendix A, we summarize the main aspects of the corrected score and unbiased estimating equations approaches. Finally, in Appendix B, we provide R scripts used to obtain the results in Chap. 5.

This book is oriented toward the needs of practitioners, laboratory scientists, applied statisticians, geostatisticians, process engineers, geologists, and analysts. It can also be a suitable book for students who specialize in the topics covered in the book.

We appreciate the support of Universidade de São Paulo, SP, Brazil (Heleno Bolfarine and Mário de Castro) and Pontificia Universidad Católica de Chile, Chile (Manuel Galea) for providing us with an adequate work environment for the development of our research and, in particular, for the conclusion of this book. Heleno Bolfarine and Mário de Castro were partially supported by CNPq and FAPESP, Brazil. Manuel Galea was partially supported by grants from CONICYT and FONDECYT, Chile. We are grateful to the Editors of the SpringerBriefs in Statistics—ABE Series (F. Louzada, H. S. Migon, G. A. Paula, and F. Cribari-Neto) for the opportunity to disseminate our work. We are indebted to Marcio V. de Castilho for introducing us to applications of the models discussed in this book in geology and in the mining industry. We would like to thank two anonymous reviewers, for their careful reading and valuable comments, which have led to a much improved version of the book. We also thank Robinson dos Santos and Saveetha Balasundaram, from Springer, for guiding us in the preparation of this book.

São Paulo, SP, Brazil — Heleno Bolfarine
São Carlos, SP, Brazil — Mário de Castro
Santiago, Chile — Manuel Galea
June 2020

# Contents

# Chapter 1
# Introduction

Comparing measuring devices, which varies in pricing, fastness, and other features such as efficiency, has been of growing interest in several areas like engineering, medicine, psychology, and agriculture (see, e.g., Meier and Zünd 2000; Woodhouse 2003; Dunn 2004; Ermer and Miller 2005; Carstensen 2010; Shoukri 2011; Lin et al. 2012; Choudhary and Nagaraja 2017). These books consider the problem of comparing measurement methods at different levels of depth, addressed topics, and type of data, continuous or discrete. The main interest is to compare different ways of measuring the same unknown quantity $x$ in a group of $n$ experimental units. We focus on the analysis of continuous data. For modeling this type of data, in the above literature it is mainly adopted the multivariate normal distribution, structural measurement error models , and mixed-effects models.

Four goals are found in the literature on the comparison of measuring methods problem (St. Laurent 1998; Dunn 2004; Stevens et al. 2017), namely, (1) calibration problems, whose objective is to establish a relationship between the measurements of new methods with the measurements of an old method; (2) comparison problems, which deal with assessing the level of agreement between two or more measurement methods whose measurements are taken on the same scale; (3) conversion problems, which deal with the comparison of measurement methods whose measurements are taken on different scales; and (4) gold-standard comparison problems, in which new methods of measurement are compared to a gold standard in order to assess the level of agreement. In this book, we focus mainly on problem (2) for continuous data.

Let $x_i$ be the true value of the unknown quantity $x$ corresponding to the experimental unit $i$ and $Y_{ij}$ the observed measurement of $x$ obtained with "new" method $j$ in the experimental unit $i$, for $i = 1, \ldots, n$ and $j = 1, \ldots, r$, where $r$ is the number of new methods to be tested. We also consider that measurements $X_i$, $i = 1, \ldots, n$, from a reference method ("old" method) are available. As in Barnett (1969) and Kimura (1992), the model we posite to relate the above quantities is given by $X_i = x_i + u_i$ and $Y_{ij} = \alpha_j + \beta_j x_i + e_{ij}$, for $i = 1, \ldots, n$ and $j = 1, \ldots, r$.



H. Bolfarine et al., *Regression Models for the Comparison of Measurement Methods*, SpringerBriefs in Statistics, https://doi.org/10.1007/978-3-030-57935-7_1

Here, $\alpha_j$ and $\beta_j$ correspond to the additive and multiplicative biases of method $j$, respectively, and quantify the biases of new measurement methods relative to the reference method, which does not present additive and multiplicative biases. This model corresponds to a multivariate measurement error model, also known as errors-in-variables model (Fuller 1987; Cheng and Van Ness 1999).

Mainly motivated by applications in geology, analytical chemistry, and data sets from the mining industry, we consider the functional version of the measurement error model, that is, we assume that $x_1, \ldots, x_n$ are incidental parameters (Cheng and Van Ness 1999, Section 1.3.2). In applications of this model, the statistical inference focuses on the additive and multiplicative biases $\alpha_j$ and $\beta_j$, respectively, for $j = 1, \ldots, r$, also known as analytical biases. As is common in these applications, we assume that the variances of the measurement errors are known and change across experimental units. Thus, for the data analysis, we assume heteroscedastic functional regression models. Even though this model has been used for data analysis in other areas (see, e.g., Walter 1997), in this book we focus on their applications in the comparison of measuring methods problem.

In what follows, we give an overview of the main topics addressed in this book, together with a discussion of some works in the literature, with emphasis on comparison of measurement methods in chemical analysis.

## 1.1 Detection of Analytical Bias

As we have already pointed out, in many cases of comparison of measuring methods the goal is to detect possible biases of the new methods. One of the hypotheses of interest to be tested is $H_0 : \alpha_j = 0$ and $\beta_j = 1$, for $j = 1, \ldots, r$, which means that the new methods measure the characteristic $x$ without biases. As stressed by de Castilho (2004), detecting analytical bias is a valuable step during method validation, and in the case of the mining industry, fundamental when one validates the geological databases used for resource estimation. This will generally impact on the costs of future investments in new or expanding projects.

Method comparison studies are the object of extensive studies in analytical chemistry, and much has been already published concerning this issue (see, e.g., Ripley and Thompson 1987; Riu and Rius 1996; Lisý et al. 1990; Martínez et al. 1999; Galea-Rojas et al. 2003; de Castro et al. 2004; de Castilho 2004; de Castro et al. 2005, 2006a,b, 2007), among many others. Riu and Rius (1996) and Martínez et al. (1999) comment on some pitfalls of the ordinary and weighted least squares approaches, and advocate the use of bivariate least squares (Lisý et al. 1990) in the construction of a joint confidence region for the additive and multiplicative biases. We find another alternative in Ripley and Thompson (1987), which assuming that the measurement errors follow Gaussian (normal) distributions, propose to estimate the additive and multiplicative biases by maximum likelihood under a functional model. These authors also present separate tests for testing additive and multiplicative biases. Galea-Rojas et al. (2003) and de Castro et al. (2004) set the problem in a functional measurement error modeling framework and under normality of the measurement errors, maximum likelihood estimation of the parameters is carried out through simple iterative steps. A Wald type statistic which guarantees correct asymptotic significance levels is proposed to test the unbiasedness of the new measurement method. de Castro et al. (2005) propose a simple pivotal quantity with exact distribution, requiring neither specialized software nor iterative procedures. For a review on techniques used in methods comparison in the mining industry, we refer to de Castilho (2004).

In this book, we set the problem in a functional measurement error modeling framework and under normality of the errors, maximum likelihood and corrected score approaches are proposed to make statistical inference on the bias parameters.

## 1.2 Probability of Agreement

Suppose two methods, labeled as new and old, furnish measurements $Y$ and $X$, respectively, of a characteristic $x$. Following Choudhary and Nagaraja (2017), agreement between the two methods refers to closeness between their measurements. Ideally, under perfect agreement we have that $P(Y = X) = 1$ and the bivariate distribution of $Y$ and $X$ is concentrated on the identity line, so that the distribution of the difference $Y - X$ is degenerate at 0. Relating these variables, as in Ripley and Thompson (1987), we write

$$Y = \alpha + \beta x + e \quad \text{and} \quad X = x + u, \tag{1.1}$$

where $e$ and $u$ are independent measurement errors with variances $\lambda$ and $\kappa$, respectively. The situation of perfect agreement is very restrictive. For example, the new method can be unbiased ($\alpha = 0$ and $\beta = 1$), nevertheless this may not be enough for good agreement because the two variances may differ considerably. Based on Eq. (1.1), we say that the two measurement methods are identical if $\alpha = 0$, $\beta = 1$, and $\lambda = \kappa$ (Stevens et al. 2017). Even in this case, with equal variances, there may not be a good agreement between the measurements of the two methods. Two measurement methods agree and could be used interchangeably if the difference in the measurements $Y - X$ is small. Typically this happens when $\alpha = 0$, $\beta = 1$, and $\lambda$ and $\kappa$ are small. For details, see Choudhary and Nagaraja (2017) and Stevens et al. (2017) and references therein.

Limits of agreement (Altman and Bland 1983; Bland and Altman 1986) have been widely used as a tool for comparing two measurement systems in many studies. Stevens et al. (2017) propose the probability of agreement as a way to overcome some shortcomings of the limits of agreement approach. For a given value $x^*$ of the true $x$, the probability of agreement is defined as $p_{\mathrm{A}}(x^*) = P(|Y - X| \leq \epsilon)$, where $\epsilon$ is the maximal practical difference, which is usually set by a specialist. Under the assumption of normality, the probability of agreement can be written as (Stevens et al. 2017)

$$p_{\mathrm{A}}(x^*) = \Phi\left(\frac{-\epsilon - \alpha - (\beta - 1)x^*}{(\lambda + \kappa)^{1/2}}\right) - \Phi\left(\frac{\epsilon - \alpha - (\beta - 1)x^*}{(\lambda + \kappa)^{1/2}}\right), \qquad (1.2)$$

where "$\Phi$" denotes the standard normal cumulative distribution function. For fixed $\epsilon$, if $\alpha = 0$ and $\beta = 1$, then the probability of agreement depends only on the standard deviation of the difference $Y - X$. Notice that, if $\lambda \to 0$ and $\kappa \to 0$ (small variances of the measurement errors), then $p_{\mathrm{A}}(x^*) \to 1$. On the contrary, if $\lambda \to \infty$ or $\kappa \to \infty$, then $p_{\mathrm{A}}(x^*) \to 0$. We use the probability of agreement to evaluate the degree of agreement between the measurements made by two methods. For a discussion on other measures of agreement, see Choudhary and Nagaraja (2017).

## 1.3 Influence Assessment

Influence assessment or sensitivity analysis is a group of techniques used to evaluate the stability of statistics of interest to perturbations in the data set and in the assumptions of a statistical model. Lee and Wang (1996) indicate that influence assessment can be seen as an extension of the robustness concept to study and detect influential subsets of data that seriously influence key results of the analysis. Thus, the detection of atypical (outliers) and influential observations is an important stage in any statistical analysis. This is essential in order to evaluate the sensitivity (robustness) of the results obtained using the available data set, since atypical observations can distort the estimators and test statistics leading to, in some cases, wrong decisions. We apply the of local influence methodology (Cook 1986; Wu and Luo 1993; Cadigan and Farrell 2002; de Castro et al. 2006b) under several perturbation schemes.

## 1.4 Our Proposal

In this book, we deal with the comparison of measurement methods problem under functional measurement error models. In our experience, this approach is suitable in many applications, especially for the detection of analytical bias in the comparison of chemical analysis methods. This background includes the explicit formulation of the functional regression model and the proposal of the statistical test for detection of biases. For parameter estimation and hypothesis testing, we use the maximum likelihood method in the presence of incidental parameters and the corrected score methodology. To the best of our knowledge, both approaches have not been considered in the existing books on this topic. Furthermore, we describe tools for the assessment of model adequacy and sensitivity analysis. Applications of the methodologies to real data sets are reported. We provide R code with functions implementing the techniques and examples so that practitioners can easily analyze their own data sets.

# Chapter 2
# Two Methods

## 2.1 Introduction

In this chapter, we consider a global procedure for testing bias absence of an alternative measurement method approach using a functional measurement error model. The approach is based on the maximum likelihood method.

Several approaches have been used with validation processes. A description of some methods is reported in Riu and Rius (1996), who advocate the bivariate least squares method (Lisý et al. 1990) and propose a confidence region for the additive and multiplicative biases. The method takes into account the uncertainty in the results that both methods may give to calculate estimators of the regression coefficients. However, the estimation of the covariance matrix is proposed in an *ad hoc* manner, and is based on results borrowed from ordinary least squares (OLS) without formal justification. For example, Riu and Rius (1996) claim that the distribution of the $F$ statistic in their expression (9) follows an $F$ distribution since it is the ratio of two chi-square distributions. But the weights in their expression are also random, for they depend on the estimator of the slope parameter. It also seems incorrect to claim independence between numerator and denominator in that expression, at least for small sample sizes. Thus, usual conditions for an $F$ distribution do not seem to be met. Furthermore, the covariance matrix estimators do not take into account errors in both axes. The main interest is to verify whether the new method is unbiased.

Another approach is proposed in Ripley and Thompson (1987). Assuming normality, maximum likelihood estimators of the additive and multiplicative biases are considered under a functional model. The estimators have to be computed numerically since no closed expressions are available for the estimators. The approach is an extension of the orthogonal least squares approach in which the measurement errors have the same variance. Ripley and Thompson (1987) also propose separate tests for testing additive and multiplicative biases, which are



H. Bolfarine et al., *Regression Models for the Comparison of Measurement Methods*, SpringerBriefs in Statistics, https://doi.org/10.1007/978-3-030-57935-7_2

also *ad hoc* extensions of the ordinary least squares approach since no consistent asymptotic covariance matrix estimators are derived. It follows that under normality, the bivariate least squares estimation approach in Riu and Rius (1996) is equivalent to the maximum likelihood approach in Ripley and Thompson (1987). However, the approaches they propose for testing statistical hypothesis involving the biases are not equivalent. Cheng and Riu (2006) describe methods based on generalized least squares estimators.

In this chapter, we consider maximum likelihood estimation of the functional regression model with variances changing according to concentration levels. Values for the variances are obtained from replicating the sample units. The asymptotic covariance matrix of the maximum likelihood estimators are derived by using general results related to properties of the maximum likelihood estimators in such models. Consistent estimators for the covariance matrix of the maximum likelihood estimators are also proposed.

The chapter is organized as follows. Section 2.2 is dedicated to model formulation. The likelihood function and an iterative approach for solving the likelihood equations are presented in Sect. 2.3. The derivation of the asymptotic covariance matrix of the parameter estimators is considered in Sect. 2.3.1. Hypothesis testing about the unbiasedness of the new measurement device is the subject of Sect. 2.3.2. Different approaches to this problem are described in Sects. 2.4–2.6. An alternative to hypothesis testing is discussed in Sect. 2.7. Results from simulation studies are discussed in Sect. 2.8. Appendix A provides short accounts of the corrected score and unbiased estimating equations approaches.

## 2.2 Model

Let $n$ be the number of samples analyzed; $X_i$, the concentration value observed by using the old method in sample $i$; $Y_i$, the concentration value observed by using the new method in sample $i$; $x_i$, the unobserved reference (true) concentration value for sample $i$, $i = 1, \ldots, n$. Throughout, vectors and matrices are typed in boldface and the notation "$\boldsymbol{t}_i \overset{\text{indep.}}{\sim} \text{N}_r(\boldsymbol{m}_i, \boldsymbol{M}_i), i = 1, \ldots, n$" means that the vectors $\boldsymbol{t}_i$ are independent and follow $r$-variate normal (Gaussian) distributions with mean vector $\boldsymbol{m}_i$ and covariance matrix $\boldsymbol{M}_i$, $i = 1, \ldots, n$.

Relating the above variables, as in Ripley and Thompson (1987), we posit the model

$$Y_i = \alpha + \beta x_i + e_i \quad \text{and} \quad X_i = x_i + u_i, \tag{2.1}$$

with the assumption that

$$\begin{pmatrix} e_i \\ u_i \end{pmatrix} \overset{\text{indep.}}{\sim} \text{N}_2\left( \begin{pmatrix} 0 \\ 0 \end{pmatrix}, \begin{bmatrix} \lambda_i & 0 \\ 0 & \kappa_i \end{bmatrix} \right). \tag{2.2}$$

Thus, with

$$\boldsymbol{Z}_i = \begin{pmatrix} X_i \\ Y_i \end{pmatrix} \quad \text{and} \quad \boldsymbol{\Sigma}_i = \begin{bmatrix} \kappa_i & 0 \\ 0 & \lambda_i \end{bmatrix},$$

we have that $\boldsymbol{Z}_i \overset{\text{indep.}}{\sim} \text{N}_2(\boldsymbol{\mu}_i, \boldsymbol{\Sigma}_i)$, with $\boldsymbol{\mu}_i = \boldsymbol{a} + x_i \boldsymbol{b}$, $\boldsymbol{a} = (0, \alpha)'$, and $\boldsymbol{b} = (1, \beta)'$, $i = 1, \ldots, n$, where the symbol "′" denotes the transpose operator. The above model is known as a functional model (Fuller 1987) since no assumption is made on the distribution of the unknown value $x_i$, $i = 1, \ldots, n$. In this situation $x_i$, $i = 1, \ldots, n$ are also unknown parameters which have to be estimated.

## 2.3 Maximum Likelihood

Letting $\boldsymbol{\theta} = (\alpha, \beta)' \in \Theta \subseteq \mathbb{R}^2$, the log-likelihood function corresponding to the model defined in (2.1) and (2.2) can be written as

$$l(\boldsymbol{\theta}) = \sum_{i=1}^{n} l_i(\boldsymbol{\theta}), \tag{2.3}$$

where

$$\begin{aligned} l_i(\boldsymbol{\theta}) &= \text{constant} - \log\{\det(\boldsymbol{\Sigma}_i)\}/2 - (\boldsymbol{Z}_i - \boldsymbol{\mu}_i)' \boldsymbol{\Sigma}_i^{-1} (\boldsymbol{Z}_i - \boldsymbol{\mu}_i)/2 \\ &= \text{constant} - \frac{1}{2} \log\{\det(\boldsymbol{\Sigma}_i)\} - \frac{1}{2} \left\{ \frac{(X_i - x_i)^2}{\kappa_i} + \frac{(Y_i - \alpha - \beta x_i)^2}{\lambda_i} \right\} \end{aligned}$$

and "$\det(\boldsymbol{\Sigma}_i)$" stands for the determinant of $\boldsymbol{\Sigma}_i$, $i = 1, \ldots, n$.

Since $x_i$, $i = 1, \ldots, n$, are unknown, to estimate $\boldsymbol{\theta}$ we consider first the profile log-likelihood function of $\boldsymbol{\theta}$. After differentiating the log-likelihood function in (2.3) with respect to $x_i$, it follows that the maximum likelihood estimator of $x_i$ as a function of $\boldsymbol{\theta}$ is given by

$$\widehat{x}_i = \frac{1}{g_i} \left\{ \frac{X_i}{\kappa_i} + \frac{\beta(Y_i - \alpha)}{\lambda_i} \right\}, \tag{2.4}$$

where

$$g_i = \frac{1}{\kappa_i} + \frac{\beta^2}{\lambda_i}, \tag{2.5}$$

$i = 1, \ldots, n$. Now, replacing $x_i$ for $\widehat{x}_i$ in (2.3), the resulting profile log-likelihood function is given by

$$l_{\mathrm{P}}(\boldsymbol{\theta}) = \sum_{i=1}^{n} h_i(\boldsymbol{\theta}), \tag{2.6}$$

where

$$h_i(\boldsymbol{\theta}) = \text{constant} - \frac{1}{2}\log\{\det(\boldsymbol{\Sigma}_i)\} - \frac{1}{2}\left\{\frac{X_i^2}{\kappa_i} + \frac{(Y_i - \alpha)^2}{\lambda_i} - g_i\widehat{x}_i^2\right\},$$

$i = 1, \ldots, n$. Maximum likelihood estimators of $\alpha$ and $\beta$ are obtained by solving the following estimating equations:

$$\frac{\partial l_{\mathrm{P}}(\boldsymbol{\theta})}{\partial \alpha} = 0 \quad \text{and} \quad \frac{\partial l_{\mathrm{P}}(\boldsymbol{\theta})}{\partial \beta} = 0,$$

which lead to the equations $\sum_{i=1}^{n}(Y_i - \alpha - \beta\widehat{x}_i)/\lambda_i = 0$ and $\sum_{i=1}^{n}(Y_i - \alpha - \beta\widehat{x}_i)\widehat{x}_i/\lambda_i = 0$. By direct computation we obtain $\mathrm{E}\{\partial l_{\mathrm{P}}(\boldsymbol{\theta})/\partial\alpha\} = 0$ and $\mathrm{E}\{\partial l_{\mathrm{P}}(\boldsymbol{\theta})/\partial\beta\} = 0$, that is, the estimating equations are unbiased (see Appendix A.2). Solving the above equations, we obtain

$$\beta = \frac{\sum_{i=1}^{n}\rho_i\widehat{x}_i(Y_i - \overline{Y}_\rho)}{\sum_{i=1}^{n}\rho_i\widehat{x}_i(\widehat{x}_i - \overline{\widehat{x}}_\rho)} \quad \text{and} \quad \alpha = \sum_{i=1}^{n}\rho_i(Y_i - \beta\widehat{x}_i), \tag{2.7}$$

where $\rho_i = \lambda^*/(n\lambda_i)$, $i = 1, \ldots, n$, $\lambda^* = n/\sum_{i=1}^{n}1/\lambda_i$, $\overline{Y}_\rho = \sum_{i=1}^{n}\rho_i Y_i$, and $\overline{\widehat{x}}_\rho = \sum_{i=1}^{n}\rho_i\widehat{x}_i$, noticing that $\sum_{i=1}^{n}\rho_i = 1$. Since $\widehat{x}_i$ depends on $\beta$, the estimators of $\alpha$ and $\beta$ have to be computed numerically. A scheme to compute estimates starts with initial values $\widehat{\alpha}^{(0)}$ and $\widehat{\beta}^{(0)}$, compute next $\widehat{x}_i$ in (2.4) and $\widehat{\beta}^{(1)}$ and $\widehat{\alpha}^{(1)}$ using (2.7), which are initial values for the next iteration and so on until convergence, which typically happens in a few iterations. Initial values may be the corrected score estimates from Sect. 2.5. Alternative approaches are discussed in Fuller (1987, Section 3.1.6) and Ripley and Thompson (1987).

### 2.3.1 Asymptotic Covariance Matrix

To estimate the asymptotic covariance matrix of the maximum likelihood estimator of $\boldsymbol{\theta}$, we apply the model-based expectation method described in Appendix A.2. To do that, we have to compute the matrices $\boldsymbol{A}$ and $\boldsymbol{V}$ in Appendix A.2, which involve the first and second derivatives of $h_i(\boldsymbol{\theta})$ with respect to $\boldsymbol{\theta}$, $i = 1, \ldots, n$. Then, defining $q_{i\alpha} = \partial h_i(\boldsymbol{\theta})/\partial\alpha$ and $q_{i\beta} = \partial h_i(\boldsymbol{\theta})/\partial\beta$, we compute, after some algebraic manipulations,

$$V = \frac{1}{n}\sum_{i=1}^{n}\begin{bmatrix}\mathrm{var}(q_{i\alpha}) & \mathrm{cov}(q_{i\alpha}, q_{i\beta})\\ \mathrm{cov}(q_{i\alpha}, q_{i\beta}) & \mathrm{var}(q_{i\beta})\end{bmatrix}$$

$$= \frac{1}{n}\sum_{i=1}^{n} w_i \begin{bmatrix}1 & x_i\\ x_i & \frac{1}{g_i}+x_i^2\end{bmatrix} = W + \begin{bmatrix}0 & 0\\ 0 & w_g\end{bmatrix},$$

where

$$w_i = \frac{1}{\lambda_i + \beta^2\kappa_i}, \quad W = \frac{1}{n}\sum_{i=1}^{n} w_i \begin{bmatrix}1 & x_i\\ x_i & x_i^2\end{bmatrix} \quad \text{and} \quad w_g = \frac{1}{n}\sum_{i=1}^{n}\frac{w_i}{g_i}. \tag{2.8}$$

Furthermore, defining $q_{i\alpha\alpha} = \partial q_{i\alpha}/\partial\alpha$, $q_{i\alpha\beta} = \partial q_{i\alpha}/\partial\beta$, and $q_{i\beta\beta} = \partial q_{i\beta}/\partial\beta$, it can be shown that $A = \mathrm{E}(-n^{-1}\ddot{L}) = -W$, where

$$\ddot{L} = \sum_{i=1}^{n}\begin{bmatrix}q_{i\alpha\alpha} & q_{i\alpha\beta}\\ q_{i\alpha\beta} & q_{i\beta\beta}\end{bmatrix}. \tag{2.9}$$

Moreover, in large samples the distribution of the maximum likelihood estimator $\widehat{\theta} = (\widehat{\alpha}, \widehat{\beta})'$ is close to the normal distribution (see Appendix A.2) with mean vector $\theta$ and approximate covariance matrix $n^{-1}\Omega$, where

$$\Omega = W^{-1}VW^{-1} = \begin{bmatrix}\sigma_{\alpha\alpha} & \sigma_{\alpha\beta}\\ \sigma_{\alpha\beta} & \sigma_{\beta\beta}\end{bmatrix}, \tag{2.10}$$

with

$$\sigma_{\alpha\alpha} = \frac{1}{\sum_{i=1}^{n} w_i} + \frac{\overline{x}_w^2}{SS_w} + nw_g\left(\frac{\overline{x}_w}{SS_w}\right)^2, \quad \sigma_{\alpha\beta} = \sigma_{\beta\alpha} = -\frac{\overline{x}_w}{SS_w}\left(1 + \frac{nw_g}{SS_w}\right)$$

$$\text{and} \quad \sigma_{\beta\beta} = \frac{1}{SS_w}\left(1 + \frac{nw_g}{SS_w}\right),$$

where $SS_w = \sum_{i=1}^{n} w_i(x_i - \overline{x}_w)^2$, $\overline{x}_w = \sum_{i=1}^{n} w_i x_i / \sum_{i=1}^{n} w_i$, $i = 1, \ldots, n$. To obtain a consistent estimator of the matrix $\Omega$, we replace $\theta$ by its consistent estimator $\widehat{\theta}$. Since $\mathrm{E}(\widehat{x}_i)=x_i$ and $\mathrm{E}(\widehat{x}_i^2 - 1/g_i)=x_i^2$, $x_i$ and $x_i^2$ are estimated by $\widehat{x}_i$ and $\widehat{x}_i^2 - 1/\widehat{g}_i$, respectively, with $g_i$ as in (2.5), $i = 1, \ldots, n$. All the expectations are calculated as in Appendix A.2, that is, they are taken with respect to the true parameter values.

### 2.3.2 Hypothesis Testing

We consider now testing the hypothesis

$$\mathrm{H}_0 : \begin{pmatrix} \alpha \\ \beta \end{pmatrix} = \begin{pmatrix} 0 \\ 1 \end{pmatrix}, \tag{2.11}$$

which corresponds to conclude that the new method is unbiased.

The Wald statistic, given by

$$W = n(\widehat{\boldsymbol{\theta}} - \boldsymbol{\theta}_0)' \widehat{\boldsymbol{\Omega}}^{-1} (\widehat{\boldsymbol{\theta}} - \boldsymbol{\theta}_0), \tag{2.12}$$

where $\widehat{\boldsymbol{\Omega}} = \boldsymbol{\Omega}$ given in (2.10) evaluated at $\boldsymbol{\theta} = \widehat{\boldsymbol{\theta}}$, is asymptotically distributed according to the $\chi_2^2$ distribution (the chi-square distribution with 2 degrees of freedom), with $\boldsymbol{\theta}_0 = (0, 1)'$. A confidence region with asymptotic confidence coefficient $\gamma$ for $\boldsymbol{\theta}$ is given by $\{\boldsymbol{\theta} \in \mathbb{R}^2 : n(\widehat{\boldsymbol{\theta}} - \boldsymbol{\theta})' \widehat{\boldsymbol{\Omega}}^{-1} (\widehat{\boldsymbol{\theta}} - \boldsymbol{\theta}) \le \chi^2_{2,\gamma}\}$, where $\chi^2_{2,\gamma}$ stands for the $\gamma$-th upper quantile of the $\chi_2^2$ distribution. The hypothesis in (2.11) is not rejected at the significance level $1 - \gamma$ if this region contains $\boldsymbol{\theta}_0 = (0, 1)'$, otherwise (2.11) is rejected.

Individual hypothesis like $\mathrm{H}_0 : \beta = 1$ can be similarly tested. For this hypothesis, we can use $Z_C = (\widehat{\beta} - 1)/\widehat{\sigma}_{\beta\beta}^{1/2}$, which for large samples is distributed according to the standard normal distribution. Ripley and Thompson (1987) suggest another statistic to test $\mathrm{H}_0 : \beta = 1$, namely, $Z_{RT} = (\widehat{\beta} - 1)/\{1/\sum_{i=1}^n \widehat{w}_i (X_i - \overline{X}_w)^2\}^{1/2}$, where $\overline{X}_w = \sum_{i=1}^n \widehat{w}_i X_i / \sum_{i=1}^n \widehat{w}_i$ and $\widehat{w}_i$ is obtained from (2.8) after replacing $\beta$ by $\widehat{\beta}$. Similarly, a separate test is proposed for testing $\mathrm{H}_0 : \alpha = 0$. Notice that the confidence region resulting from the separate tests has unknown confidence level.

## 2.4 Bivariate Least Squares

The approach in Riu and Rius (1996) (see also Sprent 1966; Mak 1983; Lisý et al. 1990; Cheng and Riu 2006) consists in estimating $\boldsymbol{\theta} = (\alpha, \beta)'$ by solving equation (4) in their paper, which can be shown to be equivalent to solving the likelihood equations in Sect. 2.3. The hypothesis $\mathrm{H}_0$ in (2.11) is tested by means of the Wald statistic in (2.12) with $n\widehat{\boldsymbol{\Omega}}^{-1}$ replaced by the matrix

$$\frac{1}{s^2} \sum_{i=1}^{n} \begin{bmatrix} \widehat{w}_i & X_i \widehat{w}_i \\ X_i \widehat{w}_i & X_i^2 \widehat{w}_i \end{bmatrix},$$

where $s^2 = \sum_{i=1}^n \widehat{w}_i (Y_i - \widehat{\alpha} - \widehat{\beta} X_i)^2/(n-2)$ and $\widehat{w}_i$ is as in Sect. 2.3.2. Riu and Rius (1996) relate the statistic to the $F$ distribution with 2 and $n - 2$ degrees of freedom, which is close to the $\chi_2^2$ distribution for moderate sample sizes.

## 2.5 Corrected Score

By using the corrected score approach (Stefanski 1989; Nakamura 1990; Gimenez and Bolfarine 1997), in this section we obtain consistent estimators for the additive and multiplicative biases in the model presented in Sect. 2.2. Furthermore, the approach yields closed form expressions for the estimators of the parameters ($\alpha$ and $\beta$). Thus, the approach is simple to implement with existing software.

In (2.1), with $e_i \overset{\text{indep.}}{\sim} \mathrm{N}(0, \lambda_i)$, $i = 1, \ldots, n$, the log-likelihood function is

$$l(\boldsymbol{\theta}; \boldsymbol{x}, \boldsymbol{Y}) = \text{constant} - \frac{1}{2}\sum_{i=1}^{n} \frac{(Y_i - \alpha - \beta x_i)^2}{\lambda_i}, \tag{2.13}$$

$i = 1, \ldots, n$, where $\boldsymbol{x} = (x_1, \ldots, x_n)'$ and $\boldsymbol{Y} = (Y_1, \ldots, Y_n)'$. Note that the above function is unobserved because it depends on the unobserved $x_i$, $i = 1, \ldots, n$. We call attention to the fact that replacing the unobserved $x_i$ by the observed $X_i$ directly in (2.13) leads to the usual (naive) weighted least squares approach, which yields inconsistent estimators of $\alpha$ and $\beta$.

Moreover, noticing that $\mathrm{E}(X_i) = x_i$ and $\mathrm{E}(X_i^2) = \kappa_i + x_i^2$, we arrive at the corrected log-likelihood function $l^*(\boldsymbol{\theta}; \boldsymbol{X}, \boldsymbol{Y}) = \sum_{i=1}^{n} l_i^*(\boldsymbol{\theta})$, where $\boldsymbol{X} = (X_1, \ldots, X_n)'$ and

$$l_i^*(\boldsymbol{\theta}) = \text{constant} - \frac{1}{2}\sum_{i=1}^{n} \frac{(Y_i - \alpha - \beta X_i)^2 - \beta^2\kappa_i}{\lambda_i}.$$

An outline of the corrected score approach is provided in Appendix A.1. Thus, maximizing $l^*(\boldsymbol{\theta}; \boldsymbol{X}, \boldsymbol{Y})$ with respect to $\alpha$ and $\beta$ will lead to the corrected score estimators of $\alpha$ and $\beta$. Now, differentiating $l^*(\boldsymbol{\theta}; \boldsymbol{X}, \boldsymbol{Y})$ with respect to $\alpha$ and $\beta$, we obtain the score function $\boldsymbol{U}^* = (U_\alpha^*, U_\beta^*)'$, where

$$U_\alpha^* = \sum_{i=1}^{n} U_{i\alpha}^* \quad \text{and} \quad U_\beta^* = \sum_{i=1}^{n} U_{i\beta}^*, \tag{2.14}$$

with $U_{i\alpha}^* = (Y_i - \alpha - \beta X_i)/\lambda_i$ and $U_{i\beta}^* = \{(Y_i - \alpha - \beta X_i)X_i + \beta\kappa_i\}/\lambda_i, i = 1, \ldots, n$. By direct computation we obtain $\mathrm{E}(U_\alpha^*) = 0$ and $\mathrm{E}(U_\beta^*) = 0$, that is, the estimating equations are unbiased (see Appendix A.2). Solving the equation $\boldsymbol{U}^* = \boldsymbol{0}_2$, where $\boldsymbol{0}_2$ denotes the bidimensional null vector, leads to the corrected score estimators, which are given by

$$\widetilde{\beta} = \frac{\sum_{i=1}^{n} \frac{X_i Y_i}{\lambda_i} - \sum_{i=1}^{n} \frac{Y_i}{\lambda_i} \sum_{i=1}^{n} \frac{X_i}{\lambda_i} (\sum_{i=1}^{n} \frac{1}{\lambda_i})^{-1}}{\sum_{i=1}^{n} \frac{X_i^2}{\lambda_i} - (\sum_{i=1}^{n} \frac{X_i}{\lambda_i})^2 (\sum_{i=1}^{n} \frac{1}{\lambda_i})^{-1} - \sum_{i=1}^{n} \frac{\kappa_i}{\lambda_i}} \quad \text{and}$$

$$\widetilde{\alpha} = \frac{\sum_{i=1}^{n}(Y_i - \widetilde{\beta}X_i)/\lambda_i}{\sum_{i=1}^{n} 1/\lambda_i}.$$

Notice that if $\kappa_i = \kappa$ and $\lambda_i = \lambda$, $i = 1, \ldots, n$, then $\widetilde{\beta} = S_{XY}/(S_X^2 - \kappa)$, which can be seen as a corrected least squares estimator where $S_{XY} = \sum_{i=1}^{n}(X_i - \overline{X})(Y_i - \overline{Y})/n$ and $S_X^2 = \sum_{i=1}^{n}(X_i - \overline{X})^2/n$, with $\overline{X} = \sum_{i=1}^{n} X_i/n$ and $\overline{Y} = \sum_{i=1}^{n} Y_i/n$. We recall that the least squares estimator of $\beta$ is given by $S_{XY}/S_X^2$. We see that the corrected score approach provides analytical expressions for the estimators of $\alpha$ and $\beta$, which is not the case with other approaches (as in Ripley and Thompson 1987; Riu and Rius 1996).

We can also obtain an estimator of the asymptotic covariance matrix of the corrected score estimator by using the sandwich method in Appendix A.2. From (2.14), simple algebraic manipulations show that

$$\frac{\partial U_\alpha^*}{\partial \alpha} = \sum_{i=1}^{n} \frac{1}{\lambda_i}, \quad \frac{\partial U_\beta^*}{\partial \beta} = \sum_{i=1}^{n} \frac{X_i^2 - \kappa_i}{\lambda_i} \quad \text{and} \quad \frac{\partial U_\alpha^*}{\partial \beta} = \frac{\partial U_\beta^*}{\partial \alpha} = \sum_{i=1}^{n} \frac{X_i}{\lambda_i},$$

$i = 1, \ldots, n$. Putting together these derivatives, we obtain the symmetric matrix $\widetilde{\boldsymbol{A}} = -n^{-1}\partial \boldsymbol{U}^*/\,\partial \boldsymbol{\theta}$, noticing that $\widetilde{\boldsymbol{A}}$ does not depend on $\alpha$ and $\beta$. Moreover, from (2.14) we also compute the elements of the matrix $\widetilde{\boldsymbol{V}} = n^{-1}\sum_{i=1}^{n} \boldsymbol{U}_i^* \boldsymbol{U}_i^{*\prime}$, where $\boldsymbol{U}_i^* = (U_{i\alpha}^*, U_{i\beta}^*)'$, $i = 1, \ldots, n$. According to Appendix A.2, an estimator of the asymptotic covariance matrix of $\widetilde{\boldsymbol{\theta}} = (\widetilde{\alpha}, \widetilde{\beta})'$ is given by $n^{-1}\widetilde{\boldsymbol{\Omega}}_{CS}$, where $\widetilde{\boldsymbol{\Omega}}_{CS} = \widetilde{\boldsymbol{A}}^{-1}\widetilde{\boldsymbol{V}}\widetilde{\boldsymbol{A}}^{-1}$ evaluated at $\boldsymbol{\theta} = \widetilde{\boldsymbol{\theta}}$. Now we can propose a Wald statistic for testing (2.11) with guaranteed asymptotic significance level. The test statistic is given by

$$W_{CS} = n(\widetilde{\boldsymbol{\theta}} - \boldsymbol{\theta}_0)'\widetilde{\boldsymbol{\Omega}}_{CS}^{-1}(\widetilde{\boldsymbol{\theta}} - \boldsymbol{\theta}_0). \tag{2.15}$$

A confidence region with asymptotic confidence coefficient $\gamma$ for $\boldsymbol{\theta}$ is given by $\{\boldsymbol{\theta} \in \mathbb{R}^2 : n(\widetilde{\boldsymbol{\theta}} - \boldsymbol{\theta})'\widetilde{\boldsymbol{\Omega}}_{CS}^{-1}(\widetilde{\boldsymbol{\theta}} - \boldsymbol{\theta}) \leq \chi_{2,\gamma}^2\}$. Similar to Sect. 2.3.2, this region can be used to test the hypothesis in (2.11).

## 2.6 Exact Test

In this section, a simple exact test statistic to detect measurement bias is proposed. The test requires neither specialized software nor iterative procedures. A careful look at the model in Sect. 2.2 enables deriving an exact test to detecting analytical bias in the new method. The proposed test has a simple form and is simple to use in the sense that it does not require computing parameter estimates.

Maximum likelihood estimation of $\alpha$ and $\beta$ through simple iterative steps, as well as a Wald type statistic to test the hypothesis in (2.11) are discussed in Sect. 2.3. On the other hand, it follows from (2.1) that $\mathrm{E}(Y_i - \alpha - \beta X_i) = 0$ and $\mathrm{var}(Y_i - \alpha -$

$\beta X_i) = \lambda_i + \beta^2\kappa_i$ so that $(Y_i - \alpha - \beta X_i)/(\lambda_i + \beta^2\kappa_i)^{1/2}$, $i = 1, \ldots, n$, follow a standard normal distribution and are independent, implying that

$$Z = n^{-1/2} \sum_{i=1}^{n} \frac{Y_i - \alpha - \beta X_i}{(\lambda_i + \beta^2\kappa_i)^{1/2}} \tag{2.16}$$

also follows a standard normal distribution. This is a key result because the distribution of Z does not depend on $\alpha$ and $\beta$ (Z is called a pivotal quantity) .

Therefore, the set of points

$$\left\{\boldsymbol{\theta} \in \mathbb{R}^2 : |Z| \leq z_{1-\gamma/2}\right\} \tag{2.17}$$

provides a $(1 - \gamma) \times 100\%$ exact confidence region for $\boldsymbol{\theta}$, where $z_{1-\gamma/2}$ is the $(1 - \gamma/2)$ upper quantile of the standard normal distribution. When $\kappa_i = \kappa$ and $\lambda_i = \lambda$, $i = 1, \ldots, n$, equality in (2.17) defines an hyperbola and the confidence region is the unbounded strip between the two branches of the hyperbola.

To test the null hypothesis in (2.11) of no biases in the new method, first we choose the significance level ($\gamma$), then we compute Z in (2.16) with $\alpha = 0$ and $\beta = 1$ resulting in

$$Z_0 = n^{-1/2} \sum_{i=1}^{n} \frac{Y_i - X_i}{(\lambda_i + \kappa_i)^{1/2}}, \tag{2.18}$$

and if $|Z_0| \leq z_{1-\gamma/2}$, we do not reject $H_0$. In an equivalent way, $H_0$ is not rejected if the point (0, 1) belongs to the region delimited by (2.17). This procedure generalizes a test proposed in Cheng and Van Ness (1999, Section 2.4.3). The confidence region in (2.17) is unbounded.

## 2.7 Probability of Agreement

Limits of agreement (Altman and Bland 1983; Bland and Altman 1986) have been widely used as a tool for comparing two measurement systems in many studies. Stevens et al. (2017) propose the probability of agreement as a way to overcome some shortcomings of the limits of agreement approach.

Omitting the observation index $i$, for measurements $Y$ and $X$ in (2.1) and a given value $x^*$ of the true $x$, the probability of agreement is defined as $p_A(x^*) = P(|Y - X| \leq \epsilon)$, where $\epsilon$ is the maximal practical difference, which is usually set by a specialist. Based on the model in Sect. 2.2, the probability of agreement can be written as (Stevens et al. 2017)

$$p_A(x^*) = \Phi\left(\frac{\epsilon - \alpha - (\beta - 1)x^*}{(\lambda + \kappa)^{1/2}}\right) - \Phi\left(\frac{-\epsilon - \alpha - (\beta - 1)x^*}{(\lambda + \kappa)^{1/2}}\right), \tag{2.19}$$

where "$\Phi$" denotes the standard normal cumulative distribution function. For fixed $\epsilon$, if $\alpha = 0$ and $\beta = 1$ (that is, if the new measurement method is unbiased), then the probability of agreement depends only on the standard deviation of the difference $Y - X$.

From (2.19), the so-called probability of agreement plot is drawn, where the estimated probability of agreement is plotted against a set of values of $x$. Estimates of $p_A(x^*)$ are computed from (2.19) with $\boldsymbol{\theta}$ replaced by $\widehat{\boldsymbol{\theta}}$ (Sect. 2.3). This plot also includes pointwise confidence intervals obtained with the delta method (Sen and Singer 1993, Theorem 3.4.5) applied to the logit transformation of $p_A(x^*)$. If the estimates of $p_A(x^*)$ are approximately constant through the selected interval for $x^*$, it is possible to summarize the information conveyed by the probability of agreement plot in a single value. Stevens et al. (2017) recommend inspecting the probability of agreement plot to assess the agreement between the two methods. According to Stevens et al. (2017), a reasonable choice would be to require $p_A(x^*) \geq 0.95$ for every $x^*$ in an interval of practical interest. If this requirement is not attained, it is recommended to analyze each estimate separately and try to find the source of disagreement.

## 2.8 Simulations

The main results in Sects. 2.3–2.5 are based on asymptotic theory. In view of this, to gauge some properties of the proposed methods in finite samples, we resort to Monte Carlo simulations. The performance of the estimators and test statistics proposed in Sects. 2.3, 2.5, and 2.6 was assessed in simulation studies carried out in Galea-Rojas et al. (2003), de Castro et al. (2006a), and de Castro et al. (2005), respectively.

Ordinary least squares (OLS), weighted least squares (WLS), maximum likelihood (Sect. 2.3), and corrected score estimators (Sect. 2.5) are compared by means of simulated bias and mean squared error. As expected, least squares estimators have the worst behavior, whereas corrected score and maximum likelihood estimators deliver the best results.

In Galea-Rojas et al. (2003), the empirical levels of five test statistics applied to the test of the null hypothesis in (2.11) are compared at a nominal level of 5%. The test statistics are labeled as OLS, WLS, RT (Sect. 2.3.2), RR (Sect. 2.4), and ML ($W$ in Sect. 2.3.2). OLS and WLS are Wald ($W$) statistics based on the ordinary and weighted least squares estimators, given by $W = \{\sum_{i=1}^n w_i^* \widehat{\alpha}^2 + 2\sum_{i=1}^n w_i^* X_i \widehat{\alpha}(\widehat{\beta}-1) + \sum_{i=1}^n w_i^* X_i^2 (\widehat{\beta}-1)^2\}/2s^2$, where $s^2 = (n-2)^{-1}\sum_{i=1}^n (Y_i - \widehat{\alpha} - \widehat{\beta} X_i)^2/w_i^*$, $\widehat{\alpha}$ and $\widehat{\beta}$ are the OLS (WLS) estimators, $w_i^* = 1$ (OLS) and $w_i^* = \lambda_i$ (WLS), $i = 1, \ldots, n$. The results in Galea-Rojas et al. (2003) indicate that the behavior of the ML statistic is markedly good even for small sample sizes ($n = 20$).

In the simulations carried out by de Castro et al. (2006a), for small samples the rejection rates of the null hypothesis in (2.11) from the test statistic in (2.15) are not so close to the nominal level. As expected, the empirical levels of this test approach 0.05 as the sample size increases. Overall, the test statistic in Sect. 2.3.2 is the one that seems to behave best in terms of closeness to the nominal significance level when compared to competitors found in the literature.

In de Castro et al. (2005), using the exact test in Sect. 2.6, rejection rates of the null hypothesis in (2.11) are close to 5%, whichever the sample size $n \in \{10, 20, 30, 50\}$. Since this an exact procedure, the discrepancies between empirical and nominal significance levels are due to the simulation process itself. When $H_0$ in (2.11) is not true, rejection rates become larger as we go farther and, not surprising, they are closer to 1.0 when the sample size increases.

# Chapter 3
# Two or More Methods

## 3.1 Introduction

As we have seen in Chap. 2, regression techniques are commonly applied in comparing two analytical methods at several concentrations and to test the biases of one method relative to others. Galea-Rojas et al. (2003) set the problem in a functional errors-in-variables modeling framework. Under normality of the errors, maximum likelihood estimation of the parameters is achieved through simple iterative steps. A Wald type statistic is proposed to test the unbiasedness of the new measurement method. Simulation studies in Galea-Rojas et al. (2003) suggest that the Wald test attains rejection rates closer to the significance levels than the BLS test in Sect. 2.4.

The joint confidence region in Riu and Rius (1996) is not directly extendable to the more than two methods comparison problem we will deal with. The same occurs with the separate tests based on Ripley and Thompson (1987). An approach that is readily extendable to the case of several new methods is the corrected score estimation technique developed in de Castro et al. (2006a).

In this chapter, based on de Castro et al. (2004), the findings of Galea-Rojas et al. (2003) are extended to a broader setup, covering the comparison of more than two measuring devices. An outline of the chapter is as follows. Section 3.2 presents the model while the likelihood function and a simple iterative scheme for solving the likelihood equations are given in Sect. 3.3. The derivation of the asymptotic covariance matrix of the parameter estimators is the subject of Sect. 3.3.1. Hypothesis testing of the unbiasedness of the new measurement devices (a global test and tests for each new method) is the subject of Sect. 3.3.2. Results from simulation studies are discussed in Sect. 3.7. Appendix A provides short accounts of the corrected score and unbiased estimating equations approaches.



H. Bolfarine et al., *Regression Models for the Comparison of Measurement Methods*, SpringerBriefs in Statistics, https://doi.org/10.1007/978-3-030-57935-7_3

## 3.2 Model

Let $n$ be the number of samples analyzed; $X_i$, the concentration value observed by using the old method in sample $i$; $Y_{ij}$, the concentration value observed by using the method $j$ in sample $i$; $x_i$, the unobserved reference (true) concentration value for sample $i$. Relating these variables, as in de Castro et al. (2004), we consider the model (an extension of the model proposed by Ripley and Thompson 1987)

$$X_i = x_i + u_i \quad \text{and} \quad Y_{ij} = \alpha_j + \beta_j x_i + e_{ij}, \tag{3.1}$$

$i = 1, \ldots, n$ and $j = 1, \ldots, r$, and $r$ is the number of new methods to be tested. This model is also adequate in situations where the main objective is to calibrate several laboratories with respect to one reference laboratory. We assume that the errors $\boldsymbol{e}_i = (e_{i1}, \ldots, e_{ir})'$ and $u_i$ are distributed as

$$\begin{pmatrix} u_i \\ \boldsymbol{e}_i \end{pmatrix} \overset{\text{indep.}}{\sim} \mathrm{N}_{r+1}\left( \begin{pmatrix} 0 \\ \boldsymbol{0}_r \end{pmatrix}, \begin{bmatrix} \kappa_i & \boldsymbol{0}_r' \\ \boldsymbol{0}_r & \boldsymbol{D}(\boldsymbol{\lambda}_i) \end{bmatrix} \right), \quad i = 1, \ldots, n, \tag{3.2}$$

where $\boldsymbol{D}(\boldsymbol{\lambda}_i)$ is the diagonal matrix with the elements of $\boldsymbol{\lambda}_i = (\lambda_{i1}, \ldots, \lambda_{ir})'$ on the main diagonal. Similarly to the model in Chap. 2, we assume that the error variances $\boldsymbol{\lambda}_i$ and $\kappa_i$ are known and greater than 0, $i = 1, \ldots, n$.

The model defined in (3.1) can be written as

$$\boldsymbol{Z}_i = \begin{pmatrix} X_i \\ \boldsymbol{Y}_i \end{pmatrix} = \begin{pmatrix} 0 \\ \boldsymbol{\alpha} \end{pmatrix} + \begin{pmatrix} 1 \\ \boldsymbol{\beta} \end{pmatrix} x_i + \begin{pmatrix} u_i \\ \boldsymbol{e}_i \end{pmatrix},$$

where $\boldsymbol{Y}_i = (Y_{i1}, \ldots, Y_{ir})'$, $i = 1, \ldots, n$, $\boldsymbol{\alpha} = (\alpha_1, \ldots, \alpha_r)'$ and $\boldsymbol{\beta} = (\beta_1, \ldots, \beta_r)'$. Then, under assumption (3.2), it follows that $\boldsymbol{Z}_i \overset{\text{indep.}}{\sim} \mathrm{N}_{r+1}(\boldsymbol{\mu}_i, \boldsymbol{\Sigma}_i)$, where

$$\boldsymbol{\mu}_i = \begin{pmatrix} x_i \\ \boldsymbol{\alpha} + \boldsymbol{\beta} x_i \end{pmatrix} \quad \text{and} \quad \boldsymbol{\Sigma}_i = \begin{bmatrix} \kappa_i & \boldsymbol{0}_r' \\ \boldsymbol{0}_r & \boldsymbol{D}(\boldsymbol{\lambda}_i) \end{bmatrix}, \quad i = 1, \ldots, n.$$

The above model is known as a functional errors-in-variables model (Fuller 1987; Cheng and Van Ness 1999). Because no assumption is made on the distribution of the unknown concentration values $x_i$, $i = 1, \ldots, n$, they are also parameters which have to be estimated. Since our main interest is in $\boldsymbol{\alpha}$ and $\boldsymbol{\beta}$, $x_1, \ldots, x_n$ are nuisance parameters.

## 3.3 Maximum Likelihood

Let $\boldsymbol{\theta} = (\boldsymbol{\alpha}', \boldsymbol{\beta}')'$, which is of dimension $2r \times 1$ and $\boldsymbol{\theta} \in \Theta \subseteq \mathbb{R}^{2r}$. The log-likelihood function corresponding to the model defined by (3.1) with assumption (3.2) can be written as

$$l(\boldsymbol{\theta}) = \sum_{i=1}^{n} l_i(\boldsymbol{\theta}), \tag{3.3}$$

where

$$\begin{aligned} l_i(\boldsymbol{\theta}) &= \text{constant} - \log\{\det(\boldsymbol{\Sigma}_i)\}/2 - (\boldsymbol{Z}_i - \boldsymbol{\mu}_i)'\boldsymbol{\Sigma}_i^{-1}(\boldsymbol{Z}_i - \boldsymbol{\mu}_i)/2 \\ &= \text{constant} - \frac{1}{2}\log\{\det(\boldsymbol{\Sigma}_i)\} - \frac{1}{2}\left\{\frac{(X_i - x_i)^2}{\kappa_i} + \sum_{j=1}^{r}\frac{(Y_{ij} - \alpha_j - \beta_j x_i)^2}{\lambda_{ij}}\right\}, \end{aligned}$$

$i = 1, \ldots, n$. To obtain the maximum likelihood estimator of $\boldsymbol{\theta}$, we consider first the maximization of the log-likelihood function with respect to $x_i$, $i = 1, \ldots, n$. After differentiating the log-likelihood function in (3.3) with respect to $x_i$, it follows that the maximum likelihood estimator of $x_i$, for fixed $\boldsymbol{\theta}$, is given by

$$\widehat{x_i} = \frac{1}{g_i}(\boldsymbol{Z}_i - \boldsymbol{a})'\boldsymbol{\Sigma}_i^{-1}\boldsymbol{b} = \frac{1}{g_i}\left\{\frac{X_i}{\kappa_i} + (\boldsymbol{Y}_i - \boldsymbol{\alpha})'\boldsymbol{D}^{-1}(\boldsymbol{\lambda}_i)\boldsymbol{\beta}\right\}, \tag{3.4}$$

where

$$g_i = \boldsymbol{b}'\boldsymbol{\Sigma}_i^{-1}\boldsymbol{b} = \frac{1}{\kappa_i} + \boldsymbol{\beta}'\boldsymbol{D}^{-1}(\boldsymbol{\lambda}_i)\boldsymbol{\beta} = \frac{1}{\kappa_i} + \sum_{j=1}^{r}\frac{\beta_j^2}{\lambda_{ij}}, \tag{3.5}$$

for $i = 1, \ldots, n$, $\boldsymbol{b} = (1, \boldsymbol{\beta}')'$ and $\boldsymbol{a} = (0, \boldsymbol{\alpha}')'$. Now, replacing $x_i$ for $\widehat{x_i}$ in (3.3), the resulting profile log-likelihood function is given by

$$l_{\mathrm{P}}(\boldsymbol{\theta}) = \sum_{i=1}^{n} h_i(\boldsymbol{\theta}), \tag{3.6}$$

where

$$h_i(\boldsymbol{\theta}) = \text{constant} - \frac{1}{2}\log\{\det(\boldsymbol{\Sigma}_i)\} - \frac{1}{2}\left\{\frac{X_i^2}{\kappa_i} + (\boldsymbol{Y}_i - \boldsymbol{\alpha})'\boldsymbol{D}^{-1}(\boldsymbol{\lambda}_i)(\boldsymbol{Y}_i - \boldsymbol{\alpha}) - g_i\widehat{x}_i^2\right\},$$

$i = 1, \ldots, n$. Differentiation of $h_i(\boldsymbol{\theta})$ with respect to $\boldsymbol{\alpha}$ and $\boldsymbol{\beta}$ leads to

$$\boldsymbol{q}_{i\alpha} = \frac{\partial h_i(\boldsymbol{\theta})}{\partial \boldsymbol{\alpha}} = \boldsymbol{D}^{-1}(\boldsymbol{\lambda}_i)(\boldsymbol{Y}_i - \boldsymbol{\alpha} - \boldsymbol{\beta}\widehat{x}_i) \tag{3.7}$$

$$\text{and} \quad \boldsymbol{q}_{i\beta} = \frac{\partial h_i(\boldsymbol{\theta})}{\partial \boldsymbol{\beta}} = \boldsymbol{D}^{-1}(\boldsymbol{\lambda}_i)(\boldsymbol{Y}_i - \boldsymbol{\alpha} - \boldsymbol{\beta}\widehat{x}_i)\widehat{x}_i, \tag{3.8}$$

$i = 1, \ldots, n$. Hence, the maximum likelihood estimators are solutions to the estimating equations

$$\frac{\partial l_{\mathrm{P}}(\boldsymbol{\theta})}{\partial \boldsymbol{\alpha}} = \mathbf{0}_r \quad \text{and} \quad \frac{\partial l_{\mathrm{P}}(\boldsymbol{\theta})}{\partial \boldsymbol{\beta}} = \mathbf{0}_r, \tag{3.9}$$

or equivalently, $\sum_{i=1}^{n} \boldsymbol{D}^{-1}(\boldsymbol{\lambda}_i)(\boldsymbol{Y}_i - \boldsymbol{\alpha} - \boldsymbol{\beta}\widehat{x}_i) = \mathbf{0}_r$ and $\sum_{i=1}^{n} \boldsymbol{D}^{-1}(\boldsymbol{\lambda}_i)(\boldsymbol{Y}_i - \boldsymbol{\alpha} - \boldsymbol{\beta}\widehat{x}_i)\widehat{x}_i = \mathbf{0}_r$. By direct computation we obtain $\mathrm{E}\{\partial l_{\mathrm{P}}(\boldsymbol{\theta})/\partial \boldsymbol{\alpha}\} = \mathbf{0}_r$ and $\mathrm{E}\{\partial l_{\mathrm{P}}(\boldsymbol{\theta})/\partial \boldsymbol{\beta}\} = \mathbf{0}_r$, that is, the estimating equations are unbiased (see Appendix A.2). Solving (3.9) yields

$$\beta_j = \frac{\sum_{i=1}^{n} \rho_{ij}\widehat{x}_i(Y_{ij} - \overline{Y}_{\rho j})}{\sum_{i=1}^{n} \rho_{ij}\widehat{x}_i(\widehat{x}_i - \overline{\widehat{x}}_{\rho j})} \quad \text{and} \quad \alpha_j = \sum_{i=1}^{n} \rho_{ij}(Y_{ij} - \beta_j\widehat{x}_i), \tag{3.10}$$

where $\rho_{ij} = \lambda_j^*/(n\lambda_{ij})$, $i = 1, \ldots, n$, $\lambda_j^* = n/\sum_{i=1}^{n} 1/\lambda_{ij}$, $\overline{Y}_{\rho j} = \sum_{i=1}^{n} \rho_{ij}Y_{ij}$, and $\overline{\widehat{x}}_{\rho j} = \sum_{i=1}^{n} \rho_{ij}\widehat{x}_i$, $j = 1, \ldots, r$, noticing that $\sum_{i=1}^{n} \rho_{ij} = 1$, $j = 1, \ldots, r$.

Maximum likelihood estimates of $\boldsymbol{\alpha}$ and $\boldsymbol{\beta}$ are obtained through iterative steps, as sketched below:

1. Let $\widehat{\boldsymbol{\alpha}}^{(0)}$ and $\widehat{\boldsymbol{\beta}}^{(0)}$ be the starting values of $\widehat{\boldsymbol{\alpha}}$ and $\widehat{\boldsymbol{\beta}}$ (for example, the corrected score estimates from Sect. 3.4). Initialize the iteration counter ($m = 0$);
2. Increment $m$ by 1;
3. Compute $\widehat{x}_i^{(m)}$ in (3.4) and $w_{ij}$ as above, $j = 1, \ldots, r$ and $i = 1, \ldots, n$;
4. Compute $\widehat{\beta}_j^{(m)}$ and $\widehat{\alpha}_j^{(m)}$ in (3.10), $j = 1, \ldots, r$;
5. Repeat steps 2–4 until convergence.

Our stopping rule is based on the maximum relative difference between estimates in two successive iterations. Steps 2–4 are repeated until

$$\max \left( \frac{\left|\widehat{\alpha}_j^{(m)} - \widehat{\alpha}_j^{(m-1)}\right|}{\left|\widehat{\alpha}_j^{(m-1)}\right|}, \frac{\left|\widehat{\beta}_j^{(m)} - \widehat{\beta}_j^{(m-1)}\right|}{\left|\widehat{\beta}_j^{(m-1)}\right|}, \ j = 1, \ldots, r \right) \tag{3.11}$$

is less than a specified tolerance. It is remarkable that these iterative steps (as well as the results in Sects. 3.3.1 and 3.3.2) can be implemented in a small piece of code using a matrix language. If $r = 1$, it can be seen that these iterative steps result in the estimates in Sect. 2.3.

The maximum likelihood estimator is a particular case of the so-called multivariate element-wise weighted total least squares (EW-TLS) estimator with right-hand side and row-wise different diagonal error covariance matrices (Kukush et al. 2002; Kukush and Van Huffel 2004; Markovsky et al. 2006). Algebraic manipulations of $l_P(\boldsymbol{\theta})$ in (3.6) lead us to the objective function of Kukush and Van Huffel (2004). The algorithm just proposed essentially takes advantage of the simpler model in this chapter.

The weighted least square cycles in the estimation of $\boldsymbol{\alpha}$ and $\boldsymbol{\beta}$ are Gauss–Seidel iterations. When the matrix $\ddot{\boldsymbol{L}}$ in (3.12) is negative definite at the solution $\widehat{\boldsymbol{\theta}}$ and the starting point is close to the solution, the Gauss–Seidel steps will converge (Thisted 1988). The convergence to the maximum of $l_P(\boldsymbol{\theta})$ in (3.6) is monotonic (Smyth 1996). Under our working conditions (see Appendix A.2), for $n$ large enough, $\ddot{\boldsymbol{L}}$ is a negative definite matrix and $\mathrm{E}(\ddot{\boldsymbol{L}})$ is close to $\ddot{\boldsymbol{L}}$ evaluated at $\boldsymbol{\theta} = \widehat{\boldsymbol{\theta}}$, so that the proposed algorithm will converge.

### 3.3.1 Asymptotic Covariance Matrix

To estimate the asymptotic covariance matrix of the maximum likelihood estimator of $\boldsymbol{\theta}$, we rely on the model-based expectation method in Appendix A.2. We have to compute the matrices $\boldsymbol{A}$ and $\boldsymbol{V}$ in Appendix A.2. First, we define the $r \times r$ matrices

$$\boldsymbol{q}_{i\alpha\alpha} = \frac{\partial^2 h_i(\boldsymbol{\theta})}{\partial \boldsymbol{\alpha} \partial \boldsymbol{\alpha}'}, \quad \boldsymbol{q}_{i\alpha\beta} = \frac{\partial^2 h_i(\boldsymbol{\theta})}{\partial \boldsymbol{\beta} \partial \boldsymbol{\alpha}'} \quad \text{and} \quad \boldsymbol{q}_{i\beta\beta} = \frac{\partial^2 h_i(\boldsymbol{\theta})}{\partial \boldsymbol{\beta} \partial \boldsymbol{\beta}'},$$

$i = 1, \ldots, n$. Let $\ddot{\boldsymbol{L}} = \partial^2 l_P(\boldsymbol{\theta}) / \partial \boldsymbol{\theta} \partial \boldsymbol{\theta}'$, so that

$$\ddot{\boldsymbol{L}} = \sum_{i=1}^{n} \begin{bmatrix} \boldsymbol{q}_{i\alpha\alpha} & \boldsymbol{q}_{i\alpha\beta} \\ \boldsymbol{q}'_{i\alpha\beta} & \boldsymbol{q}_{i\beta\beta} \end{bmatrix}. \tag{3.12}$$

Next, from (3.7) and (3.8) we obtain $\boldsymbol{q}_{i\alpha\alpha} = -\boldsymbol{P}_i$,

$$\boldsymbol{q}_{i\alpha\beta} = -\boldsymbol{D}^{-1}(\boldsymbol{\lambda}_i) \left\{ \frac{1}{g_i} \boldsymbol{\beta} (\boldsymbol{Y}_i - \boldsymbol{\alpha} - 2\widehat{x}_i \boldsymbol{\beta})' \boldsymbol{D}^{-1}(\boldsymbol{\lambda}_i) + \widehat{x}_i \boldsymbol{I}_r \right\},$$

and $$\boldsymbol{q}_{i\beta\beta} = \boldsymbol{D}^{-1}(\boldsymbol{\lambda}_i) \left\{ \frac{1}{g_i} (\boldsymbol{Y}_i - \boldsymbol{\alpha} - 2\boldsymbol{\beta}\widehat{x}_i)(\boldsymbol{Y}_i - \boldsymbol{\alpha} - 2\widehat{x}_i \boldsymbol{\beta})' \boldsymbol{D}^{-1}(\boldsymbol{\lambda}_i) - \widehat{x}_i^2 \boldsymbol{I}_r \right\},$$

where

$$\boldsymbol{P}_i = \boldsymbol{D}^{-1}(\boldsymbol{\lambda}_i) \left\{ \boldsymbol{I}_r - \frac{1}{g_i} \boldsymbol{\beta}\boldsymbol{\beta}' \boldsymbol{D}^{-1}(\boldsymbol{\lambda}_i) \right\} = \left\{ \boldsymbol{D}(\boldsymbol{\lambda}_i) + \kappa_i \boldsymbol{\beta}\boldsymbol{\beta}' \right\}^{-1},$$

$g_i$ comes from (3.5), $i = 1, \ldots, n$, and $\boldsymbol{I}_r$ denotes the unity matrix of order $r$. Thus, we obtain

$$\boldsymbol{A} = \mathrm{E}(-n^{-1}\ddot{\boldsymbol{L}}) = \frac{1}{n}\sum_{i=1}^{n}\begin{bmatrix} 1 & x_i \\ x_i & x_i^2 \end{bmatrix} \otimes \boldsymbol{P}_i,$$

where "⊗" denotes the Kronecker product between matrices.

After computing covariances, we arrive at

$$\begin{aligned}\boldsymbol{V} &= \frac{1}{n}\sum_{i=1}^{n}\begin{bmatrix} \mathrm{cov}(\boldsymbol{q}_{i\alpha}) & \mathrm{cov}(\boldsymbol{q}_{i\alpha}, \boldsymbol{q}_{i\beta}) \\ \mathrm{cov}(\boldsymbol{q}_{i\alpha}, \boldsymbol{q}_{i\beta}) & \mathrm{cov}(\boldsymbol{q}_{i\beta}) \end{bmatrix} \\ &= \frac{1}{n}\sum_{i=1}^{n}\begin{bmatrix} 1 & x_i \\ x_i & \frac{1}{g_i} + x_i^2 \end{bmatrix} \otimes \boldsymbol{P}_i = \boldsymbol{A} + \boldsymbol{F}, \end{aligned} \qquad (3.13)$$

where

$$\boldsymbol{F} = \begin{bmatrix} \boldsymbol{0}_{r,r} & \boldsymbol{0}_{r,r} \\ \boldsymbol{0}_{r,r} & \frac{1}{n}\sum_{i=1}^{n}\frac{1}{g_i}\boldsymbol{P}_i \end{bmatrix}.$$

Moreover, in large samples the distribution of $\widehat{\boldsymbol{\theta}}$ is close to the normal distribution (see Appendix A.2) with mean vector $\boldsymbol{\theta}$ and approximate covariance matrix $\mathrm{cov}(\widehat{\boldsymbol{\theta}}) = n^{-1}\boldsymbol{\Omega}$, where

$$\boldsymbol{\Omega} = \boldsymbol{A}^{-1}\boldsymbol{V}\boldsymbol{A}^{-1}. \qquad (3.14)$$

To obtain a consistent estimator of the matrix $\boldsymbol{\Omega}$, we replace $\boldsymbol{\theta}$ by its consistent estimator $\widehat{\boldsymbol{\theta}}$. Since $\mathrm{E}(\widehat{x}_i)=x_i$ and $\mathrm{E}(\widehat{x}_i^2 - 1/g_i)=x_i^2$, $x_i$ and $x_i^2$ can be estimated by $\widehat{x}_i$ and $\widehat{x}_i^2 - 1/\widehat{g}_i$, respectively, with $g_i$ as in (3.5), $i = 1, \ldots, n$.

From (3.13) and (3.14), we obtain

$$\mathrm{cov}(\widehat{\boldsymbol{\theta}}) = n^{-1}\boldsymbol{A}^{-1} + n^{-1}\boldsymbol{A}^{-1}\boldsymbol{F}\boldsymbol{A}^{-1}. \qquad (3.15)$$

The second term in (3.15) is a positive semi-definite matrix. Hence, the asymptotic variances of the estimators $\widehat{\boldsymbol{\alpha}}$ and $\widehat{\boldsymbol{\beta}}$ can exceed the Cramér–Rao lower bound given by the first term in (3.15). All the above expectations are calculated as in Appendix A.2, that is, they are taken with respect to the true parameter values.

### 3.3.2 Hypothesis Testing

When dealing with the methods comparison problem, it is of interest to test the hypothesis

$$\mathrm{H}_0 : \begin{pmatrix}\boldsymbol{\alpha}\\ \boldsymbol{\beta}\end{pmatrix} = \begin{pmatrix}\mathbf{0}_r\\ \mathbf{1}_r\end{pmatrix}, \tag{3.16}$$

where "$\mathbf{1}_r$" denotes the $r$-dimensional vector of ones. which means no biases in the new methods. Furthermore, one would like to carry out separate tests for each new method. However, one would want to use all the available data because all the measuring devices provide information about the true $x$ values. In this case, we are interested in testing

$$\mathrm{H}_{0j} : \begin{pmatrix}\alpha_j\\ \beta_j\end{pmatrix} = \begin{pmatrix}0\\ 1\end{pmatrix}, \quad j \in \{1, \ldots, r\}. \tag{3.17}$$

The hypothesis in (3.16) can be tested using the Wald statistic in (2.12), with $\boldsymbol{\Omega}$ given in (3.14) and $\boldsymbol{\theta}_0 = (\mathbf{0}_r', \mathbf{1}_r')'$, which is asymptotically distributed according to the $\chi^2_{2r}$ distribution. Likewise, we can test the hypothesis in (3.17) with the statistic

$$W_j = n(\widehat{\boldsymbol{\theta}}_j - \boldsymbol{\theta}_0^*)' \begin{bmatrix} \widehat{\varpi}^{j,j} & \widehat{\varpi}^{j,j+p} \\ \widehat{\varpi}^{j,j+p} & \widehat{\varpi}^{j+p,j+p} \end{bmatrix} (\widehat{\boldsymbol{\theta}}_j - \boldsymbol{\theta}_0^*), \tag{3.18}$$

where $\widehat{\boldsymbol{\theta}}_j = (\widehat{\alpha}_j, \widehat{\beta}_j)'$, $\boldsymbol{\theta}_0^* = (0, 1)'$ and $\widehat{\varpi}^{l,m}$ denotes the $(l, m)$-entry in $\boldsymbol{\Omega}^{-1}$ evaluated at $\boldsymbol{\theta} = \widehat{\boldsymbol{\theta}}$, $j \in \{1, \ldots, r\}$. As $n \to \infty$, the distribution of $W_j$ tends to the $\chi^2_2$ distribution.

## 3.4 Corrected Score

An outline of the corrected score approach can be found in Appendix A.1. Under the model in Sect. 3.2, the unobserved log-likelihood function (see Sect. 2.5) can be written as

$$l(\boldsymbol{\theta}; \boldsymbol{x}, \boldsymbol{Y}) = \text{constant} - \frac{1}{2}\sum_{i=1}^{n}\sum_{j=1}^{r}\frac{(Y_{ij} - \alpha_j - \beta_j x_i)^2}{\lambda_{ij}},$$

where $\boldsymbol{Y} = (\boldsymbol{Y}_1', \ldots, \boldsymbol{Y}_n')'$. Since $\mathrm{E}(X_i) = x_i$ and $\mathrm{E}(X_i^2) = \kappa_i + x_i^2$, $i = 1, \ldots, n$, after some algebraic manipulations it follows that the corrected log-likelihood function is given by $l^*(\boldsymbol{\theta}; \boldsymbol{X}, \boldsymbol{Y}) = \sum_{i=1}^{n} l_i^*(\boldsymbol{\theta})$, where

$$l_i^*(\boldsymbol{\theta}) = \text{constant} - \frac{1}{2}\sum_{i=1}^{n}\sum_{j=1}^{r}\frac{(Y_{ij}-\alpha_j-\beta_j X_i)^2-\beta_j^2\kappa_i}{\lambda_{ij}}.$$

Differentiating $l^*(\boldsymbol{\theta}; \boldsymbol{X}, \boldsymbol{Y})$ with respect to $\boldsymbol{\alpha}$ and $\boldsymbol{\beta}$, we obtain the score function $\boldsymbol{U}^* = (\boldsymbol{U}_\alpha^{*'}, \boldsymbol{U}_\beta^{*'})'$, where

$$U_{\alpha j}^* = \sum_{i=1}^{n} U_{i\alpha j}^* \quad \text{and} \quad U_{\beta j}^* = \sum_{i=1}^{n} U_{i\beta j}^*, \tag{3.19}$$

with $U_{i\alpha j}^* = (Y_{ij}-\alpha_j-\beta_j X_i)/\lambda_{ij}$ and $U_{i\beta j}^* = \{(Y_{ij}-\alpha_j-\beta_j X_i)X_i + \beta_j\kappa_i\}/\lambda_{ij}$, $j = 1,\ldots,r$ and $i = 1,\ldots,n$. By direct computation we obtain $\mathrm{E}(\boldsymbol{U}_\alpha^*) = \boldsymbol{0}_r$ and $\mathrm{E}(\boldsymbol{U}_\beta^*) = \boldsymbol{0}_r$, that is, the estimating equations are unbiased (see Appendix A.2). Solving the equation $\boldsymbol{U}^* = \boldsymbol{0}_{2r}$ yields the corrected score estimators

$$\widetilde{\beta}_j = \frac{\sum_{i=1}^n \frac{Y_{ij}X_i}{\lambda_{ij}} - \sum_{i=1}^n \frac{X_i}{\lambda_{ij}}\sum_{i=1}^n \frac{Y_{ij}}{\lambda_{ij}}(\sum_{i=1}^n \frac{1}{\lambda_{ij}})^{-1}}{\sum_{i=1}^n \frac{X_i^2}{\lambda_{ij}} - (\sum_{i=1}^n \frac{X_i}{\lambda_{ij}})^2(\sum_{i=1}^n \frac{1}{\lambda_{ij}})^{-1} - \sum_{i=1}^n \frac{\kappa_i}{\lambda_{ij}}}$$

$$\text{and} \quad \widetilde{\alpha}_j = \frac{\sum_{i=1}^n \frac{Y_{ij}-\widetilde{\beta}_j X_i}{\lambda_{ij}}}{\sum_{i=1}^n \frac{1}{\lambda_{ij}}},$$

$j = 1,\ldots,r$.

Now, to obtain an estimator of the asymptotic covariance matrix of the corrected score estimator $\widetilde{\boldsymbol{\theta}} = (\widetilde{\boldsymbol{\alpha}}', \widetilde{\boldsymbol{\beta}}')'$, we use the sandwich method in Appendix A.2. First we compute the second derivatives

$$\frac{\partial U_{\alpha j}^*}{\partial \alpha_j} = \sum_{i=1}^n \frac{1}{\lambda_{ij}}, \quad \frac{\partial U_{\beta j}^*}{\partial \beta_j} = \sum_{i=1}^n \frac{X_i^2-\kappa_i}{\lambda_{ij}}, \quad \frac{\partial U_{\alpha j}^*}{\partial \beta_j} = \frac{\partial U_{\beta j}^*}{\partial \alpha_j} = \sum_{i=1}^n \frac{X_i}{\lambda_{ij}}$$

$$\text{and} \quad \frac{\partial U_{\alpha j}^*}{\partial \alpha_k} = \frac{\partial U_{\beta j}^*}{\partial \beta_k} = \frac{\partial U_{\alpha j}^*}{\partial \beta_k} = \frac{\partial U_{\beta j}^*}{\partial \alpha_k} = 0, \quad \text{if } j \neq k,$$

$j, k = 1,\ldots,r$. The symmetric matrix $\widetilde{\boldsymbol{A}} = -n^{-1}\partial \boldsymbol{U}^*/\partial\boldsymbol{\theta}$, which is of dimension $2r \times 2r$, is such that

$$\widetilde{\boldsymbol{A}} = \begin{bmatrix} \widetilde{\boldsymbol{A}}_{\alpha\alpha} & \widetilde{\boldsymbol{A}}_{\alpha\beta} \\ \widetilde{\boldsymbol{A}}_{\alpha\beta} & \widetilde{\boldsymbol{A}}_{\beta\beta} \end{bmatrix},$$

where $\widetilde{\boldsymbol{A}}_{\alpha\alpha}$, $\widetilde{\boldsymbol{A}}_{\alpha\beta}$, and $\widetilde{\boldsymbol{A}}_{\beta\beta}$ are diagonal matrices with elements $-n^{-1}\sum_{i=1}^n 1/\lambda_{ij}$, $-n^{-1}\sum_{i=1}^n X_i/\lambda_{ij}$, and $-n^{-1}\sum_{i=1}^n (X_i^2-\kappa_i)/\lambda_{ij}$ on the main diagonal, respectively, $j = 1,\ldots,r$. Notice that $\widetilde{\boldsymbol{A}}$ does not depend on $\boldsymbol{\theta}$. Furthermore, letting

$\boldsymbol{U}_i^* = (U_{\alpha 1}^*, \ldots, U_{\alpha r}^*, U_{\beta 1}^*, \ldots, U_{\beta r}^*)'$, from (3.19) we also compute the elements of the matrix $\widetilde{\boldsymbol{V}} = \sum_{i=1}^n \boldsymbol{U}_i^* \boldsymbol{U}_i{}^{*\prime}/n$. According to Appendix A.2, an estimator of the asymptotic covariance matrix of $\widetilde{\boldsymbol{\theta}}$ is given by $n^{-1}\widetilde{\boldsymbol{\Omega}}_{CS}$, where $\widetilde{\boldsymbol{\Omega}}_{CS} = \widetilde{\boldsymbol{A}}^{-1}\widetilde{\boldsymbol{V}}\widetilde{\boldsymbol{A}}^{-1}$ evaluated at $\boldsymbol{\theta} = \widetilde{\boldsymbol{\theta}}$. The hypothesis in (3.16) can be tested using the Wald statistic in (2.15) with $\boldsymbol{\theta}_0 = (\boldsymbol{0}_r', \boldsymbol{1}_r')'$, which is asymptotically distributed according to the $\chi_{2r}^2$ distribution.

## 3.5 Exact Test

To test the null hypothesis in (3.16), the exact test statistic in Sect. 2.6 is easily generalized. Based on the model in Sect. 3.2, the statistic in (2.18) is written as

$$Z_0 = (nr)^{-1/2} \sum_{i=1}^{n} \sum_{j=1}^{r} \frac{Y_{ij} - X_i}{(\lambda_{ij} + \kappa_i)^{1/2}}. \tag{3.20}$$

## 3.6 Probability of Agreement

The probability of agreement (Stevens et al. 2017) is described in Sect. 2.7. Since the model in Sect. 3.2 comprises $r$ new measurement methods, omitting the observation index $i$ the probability of agreement in (2.19) is recast as

$$p_{\mathrm{A}}(x^*) = \Phi\left(\frac{\epsilon - \alpha_j - (\beta_j - 1)x^*}{(\lambda_j + \kappa)^{1/2}}\right) - \Phi\left(\frac{-\epsilon - \alpha_j - (\beta_j - 1)x^*}{(\lambda_j + \kappa)^{1/2}}\right), \tag{3.21}$$

for $j \in \{1, \ldots, r\}$. From (3.21), a probability of agreement plot is drawn for each new measurement method. Recall that $\epsilon$ denotes the maximal practical difference.

## 3.7 Simulations

In order to state our main results in Sect. 3.3.2, we rely on asymptotic theory. In view of this, we planned Monte Carlo simulations to evaluate the empirical level and the power of the Wald test statistic in (2.12) with $\boldsymbol{\Omega}$ in (3.14). Results in de Castro et al. (2004) indicate that, even in small samples, in general empirical and nominal levels are close, suggesting a good agreement between empirical and theoretical distributions. Moreover, when data are generated with true values of $\boldsymbol{\theta}$ going way from $\boldsymbol{\theta}_0$ in (3.16), rejection rates of $\mathrm{H}_0$ in (3.16) increase, as expected.

# Chapter 4
# Model Checking and Influence Assessment

## 4.1 Introduction

It is currently recognized that to assess the influence of data and assumptions underlying the working model on the results plays a key role in a statistical analysis. In this chapter, we address the effects of minor perturbations of data on test statistics.

A perturbation scheme is a key concept and can be understood as a way of modifying the working model. Precisely, a perturbation scheme has to be translated into a $n^*$-dimensional vector $\boldsymbol{\omega}$ restricted to some open subset of $\mathbb{R}^{n^*}$ ($n^*$ does not necessarily matches the sample size). After perturbing model components and/or data with $\boldsymbol{\omega}$, we assess the influence of these perturbations on model results. There are three major points. First, meaningful perturbation schemes should be chosen in advance. Second, a selection of which particular aspects (for example, parameter estimates or some function of them) will be assessed under the perturbations. At last, an objective criterion to quantify the impacts of perturbations has to be chosen.

This chapter is organized as follows. A device for model checking is described in Sect. 4.2. In Sect. 4.3, we provide a short account of the assessment methodologies proposed by Cook (1986), Wu and Luo (1993), and Cadigan and Farrell (2002). Some meaningful perturbation schemes applied to the models in Sects. 2.2 and 3.2 are detailed in Sects. 4.4 and 4.5, respectively.

## 4.2 Model Checking

Model checking has received much less attention than inference in the measurement error models literature. In the sequel, we describe a simple graphical device for assessing goodness of fit. Under the models in Sects. 2.2 and 3.2, we have that the



H. Bolfarine et al., *Regression Models for the Comparison of Measurement Methods*, SpringerBriefs in Statistics, https://doi.org/10.1007/978-3-030-57935-7_4

measurements $\boldsymbol{Z}_i$ are such that $\boldsymbol{Z}_i \overset{\text{indep.}}{\sim} \mathrm{N}_2(\boldsymbol{\mu}_i, \boldsymbol{\Sigma}_i)$ and $\boldsymbol{Z}_i \overset{\text{indep.}}{\sim} \mathrm{N}_{r+1}(\boldsymbol{\mu}_i, \boldsymbol{\Sigma}_i)$, $i = 1, \ldots, n$, respectively. Using the result in Sect. 2.6, the random variables $\xi_{ij} = (Y_{ij} - \alpha_j - \beta_j X_i)/(\lambda_{ij} + \beta_j^2 \kappa_i)^{1/2}$, for $j = 1, \ldots, r$ and $i = 1, \ldots, n$, follow the standard normal distribution and are independent. Therefore, except for the uncertainty due to estimation of $\boldsymbol{\theta}$, the residuals $\widehat{\xi}_{ij}$, for $j = 1, \ldots, r$ and $i = 1, \ldots, n$, follow the N(0, 1) distribution, where $\widehat{\xi}_{ij}$ is computed with $\boldsymbol{\theta}$ replaced by $\widehat{\boldsymbol{\theta}}$ from Sect. 2.2 or 3.2. This result enables us to check goodness of fit by employing the simulated envelope proposed in Atkinson (1985). The residuals $\widehat{\xi}_{ij}$, for $j = 1, \ldots, r$ and $i = 1, \ldots, n$, are ordered as $\widehat{\xi}_{(1)} < \cdots < \widehat{\xi}_{(nr)}$. First we simulate $M$ samples of $\boldsymbol{Z}_1, \ldots, \boldsymbol{Z}_n$ with $\boldsymbol{\theta} = \widehat{\boldsymbol{\theta}}$. For the $m$-th simulated sample, we compute the estimate $\breve{\boldsymbol{\theta}}$, say, and the residuals $\breve{\xi}_{mij}$, for $j = 1, \ldots, r$ and $i = 1, \ldots, n$, which are ordered as $\breve{\xi}_{m(1)} < \cdots < \breve{\xi}_{m(nr)}$. The pairs $(\Phi^{-1}\{(i-3/8)/(nr+1/4)\}, \widehat{\xi}_{(i)})$, $i = 1, \ldots, nr$, are drawn in a graph, where "$\Phi^{-1}$" denotes the quantile function of the N(0, 1) distribution. The limits of the envelope, given by $\min(\breve{\xi}_{1(i)}, \ldots, \breve{\xi}_{M(i)})$ and $\max(\breve{\xi}_{1(i)}, \ldots, \breve{\xi}_{M(i)})$, and the line connecting the points $(\Phi^{-1}\{(i - 3/8)/(nr + 1/4)\}, \sum_{m=1}^{M} \breve{\xi}_{m(i)}/M)$, $i = 1, \ldots, nr$, are also drawn in the graph. This graphical diagnostics can reveal model inadequacy.

## 4.3 Local Influence

Initially, the profile log-likelihood function in (2.6) or (3.6) is perturbed so that it takes the form $l_{\mathrm{P}}(\boldsymbol{\theta}, \boldsymbol{\omega})$. Denoting the vector of no perturbation by $\boldsymbol{\omega}_0$, we assume that $l_{\mathrm{P}}(\boldsymbol{\theta}, \boldsymbol{\omega}_0) = l_{\mathrm{P}}(\boldsymbol{\theta})$. To assess the influence of the perturbations on the estimator $\widehat{\boldsymbol{\theta}}$, the discrepancy between the two models is quantified through the likelihood displacement $LD(\boldsymbol{\omega}) = 2\{l_{\mathrm{P}}(\widehat{\boldsymbol{\theta}}) - l_{\mathrm{P}}(\widehat{\boldsymbol{\theta}}_\omega)\}$, where $\widehat{\boldsymbol{\theta}}_\omega$ denotes the maximum likelihood estimator of $\boldsymbol{\theta}$ under the perturbed model, that is, $\widehat{\boldsymbol{\theta}}_\omega$ maximizes $l_{\mathrm{P}}(\boldsymbol{\theta}, \boldsymbol{\omega})$. The idea of local influence (Cook 1986, 1987) is concerned in characterizing the behavior of $LD(\boldsymbol{\omega})$ at $\boldsymbol{\omega} = \boldsymbol{\omega}_0$. With "$||\cdot||$" denoting the Euclidean norm, the procedure consists in selecting a direction $\boldsymbol{d}$ such that $||\boldsymbol{d}|| = 1$ and then to consider the plot of $LD\big(\boldsymbol{\omega}(a)\big)$ against $a$, with $\boldsymbol{\omega}(a) = \boldsymbol{\omega}_0 + a\boldsymbol{d}$, $a \in \mathbb{R}$. This plot is called lifted line. Each lifted line can be characterized by its normal curvature $C_{\boldsymbol{d}}$ around $a = 0$. The suggestion is to pick the direction $\boldsymbol{d}_{\max}$ corresponding to the largest curvature $C_{\max}$. The normal curvature at the direction $\boldsymbol{d}$ takes the form $C_{\boldsymbol{d}} = 2|\boldsymbol{d}'\boldsymbol{\Delta}'\ddot{\boldsymbol{L}}^{-1}\boldsymbol{\Delta}\boldsymbol{d}|$ (Cook 1986), where

$$\boldsymbol{\Delta} = \begin{bmatrix} \boldsymbol{\Delta}_\alpha \\ \boldsymbol{\Delta}_\beta \end{bmatrix} = \begin{bmatrix} \dfrac{\partial^2 l_{\mathrm{P}}(\boldsymbol{\theta}, \boldsymbol{\omega})}{\partial \boldsymbol{\alpha} \partial \omega_1} & \cdots & \dfrac{\partial^2 l_{\mathrm{P}}(\boldsymbol{\theta}, \boldsymbol{\omega})}{\partial \boldsymbol{\alpha} \partial \omega_{n^*}} \\ \dfrac{\partial^2 l_{\mathrm{P}}(\boldsymbol{\theta}, \boldsymbol{\omega})}{\partial \boldsymbol{\beta} \partial \omega_1} & \cdots & \dfrac{\partial^2 l_{\mathrm{P}}(\boldsymbol{\theta}, \boldsymbol{\omega})}{\partial \boldsymbol{\beta} \partial \omega_{n^*}} \end{bmatrix} \tag{4.1}$$

and $\ddot{\boldsymbol{L}}$ as in (2.9) or (3.12), evaluated at $\boldsymbol{\theta} = \widehat{\boldsymbol{\theta}}$ and $\boldsymbol{\omega} = \boldsymbol{\omega}_0$. Notice that in Chap. 2, the vectors $\boldsymbol{\alpha}$ and $\boldsymbol{\beta}$ are scalar quantities. Maximization of $C_d$ is equivalent to finding the largest absolute eigenvalue $C_{\max}$ of the matrix $\boldsymbol{B} = \boldsymbol{\Delta}'\ddot{\boldsymbol{L}}^{-1}\boldsymbol{\Delta}$ and $\boldsymbol{d}_{\max}$ is the associated eigenvector.

It may be also of interest to assess the influence on each parameter ($\boldsymbol{\alpha}$ or $\boldsymbol{\beta}$) at time. It can be shown that the curvature at the direction $\boldsymbol{d}$ is given by $C_d^* = 2|\boldsymbol{d}'\boldsymbol{\Delta}'(\ddot{\boldsymbol{L}}^{-1} - \boldsymbol{B}_s)\boldsymbol{\Delta}\boldsymbol{d}|$, where

$$\boldsymbol{B}_s = \begin{bmatrix} \boldsymbol{0}_{r,r} & \boldsymbol{0}_{r,r} \\ \boldsymbol{0}_{r,r} & \left\{\dfrac{\partial^2 l_{\mathrm{P}}(\boldsymbol{\theta})}{\partial\boldsymbol{\beta}\partial\boldsymbol{\beta}'}\right\}^{-1} \end{bmatrix},$$

if the interest is on $\boldsymbol{\alpha}$ and

$$\boldsymbol{B}_s = \begin{bmatrix} \left\{\dfrac{\partial^2 l_{\mathrm{P}}(\boldsymbol{\theta})}{\partial\boldsymbol{\alpha}\partial\boldsymbol{\alpha}'}\right\}^{-1} & \boldsymbol{0}_{r,r} \\ \boldsymbol{0}_{r,r} & \boldsymbol{0}_{r,r} \end{bmatrix},$$

if the focus is on $\boldsymbol{\beta}$. The eigenvector $\boldsymbol{d}_{\max}$ corresponds to the largest absolute eigenvalue of the matrix $\boldsymbol{\Delta}'(\ddot{\boldsymbol{L}}^{-1} - \boldsymbol{B}_s)\boldsymbol{\Delta}$.

A direction with nonnull elements only in outstanding positions of $\boldsymbol{d}_{\max}$ is also of interest. Other important direction, according to Escobar and Meeker (1992) (see also Verbeke and Molenberghs 2000) is $\boldsymbol{v}_{jn^*} = (0, \ldots, 1, \ldots, 0)'$, the $j$-th unit direction in $\mathbb{R}^{n^*}$. In this case, the normal curvature, called the total local influence of observation $j$, reduces to $C_{dj} = 2|\boldsymbol{v}'_{jn^*}\boldsymbol{B}\boldsymbol{v}_{jn^*}| = 2|b_{jj}|$, where $b_{jj}$ is the $j$-th element on the diagonal of $\boldsymbol{B}$, $j = 1, \ldots, n^*$. As stated by Verbeke and Molenberghs (2000), the $j$-th element is influential if $C_{dj}$ is larger than the cutoff value $2\sum_{j=1}^{n^*} C_{dj}/n^*$. We use $\boldsymbol{d}_{\max}$ and $C_{dj}$, $j = 1, \ldots, n^*$, as diagnostic tools for local influence.

### 4.3.1 First-Order Approach

Although Cook's approach has proved to be useful in many applications, there are situations in that the main concerns are not on the estimator $\widehat{\boldsymbol{\theta}}$ itself, but on a test statistic $T$, like the Wald statistic in (2.12) or the $Z_0$ statistic in (2.18). Let $T_\omega(\widehat{\boldsymbol{\theta}}_\omega)$ denote the perturbed statistic, noticing that the dependence on $\boldsymbol{\omega}$ is not only through $\widehat{\boldsymbol{\theta}}_\omega$. According to the so-called first-order approach (Wu and Luo 1993; Cadigan and Farrell 2002), local influence of a perturbation is quantified by the slope $S(\boldsymbol{d})$ in direction $\boldsymbol{d}$ of the graph of $T_\omega(\widehat{\boldsymbol{\theta}}_\omega)$ versus $\boldsymbol{\omega}(a)$, given by $S(\boldsymbol{d}) = \partial T_\omega(\widehat{\boldsymbol{\theta}}_\omega)/\partial a = \boldsymbol{d}'\partial T_\omega(\widehat{\boldsymbol{\theta}}_\omega)/\partial\boldsymbol{\omega}$, evaluated at $a = 0$. Taking into account some algebraic manipulations (Cadigan and Farrell 2002), we have that

$$S(\boldsymbol{d}) = \boldsymbol{d}' \left\{ \left. \frac{\partial T_\omega(\widehat{\boldsymbol{\theta}})}{\partial \boldsymbol{\omega}} \right|_{\boldsymbol{\omega}=\boldsymbol{\omega}_0} + \left. \frac{\partial \widehat{\boldsymbol{\theta}}_\omega}{\partial \boldsymbol{\omega}}' \right|_{\boldsymbol{\omega}=\boldsymbol{\omega}_0} \times \left. \frac{\partial T(\boldsymbol{\theta})}{\partial \boldsymbol{\theta}} \right|_{\boldsymbol{\theta}=\widehat{\boldsymbol{\theta}}} \right\}. \tag{4.2}$$

For the Wald statistic in (2.12), we have $T_\omega(\widehat{\boldsymbol{\theta}}) = T(\widehat{\boldsymbol{\theta}})$, so that the first summand in (4.2) vanishes. Moreover, using results in Cadigan and Farrell (2002) we obtain $S(\boldsymbol{d}) = -\boldsymbol{d}'\dot{\boldsymbol{T}}_0$, where $\dot{\boldsymbol{T}}_0 = \boldsymbol{\Delta}'\ddot{\boldsymbol{L}}^{-1}\partial T(\boldsymbol{\theta})/\partial\boldsymbol{\theta}$ evaluated at $\boldsymbol{\theta} = \widehat{\boldsymbol{\theta}}$, with $\boldsymbol{\Delta}$ as in (4.1) and $\ddot{\boldsymbol{L}}$ as in (2.9) or (3.12). Because partial derivatives of $W$ in (2.12) are not simple and numerical differentiation is fast and reliable, we propose computing $\partial T(\boldsymbol{\theta})/\partial\boldsymbol{\theta}$ numerically.

The pivotal quantity statistic in (2.18) does not depend on $\boldsymbol{\theta}$. Thus, $T_\omega(\widehat{\boldsymbol{\theta}}) = T_\omega$ and the second summand in (4.2) vanishes so that $S(\boldsymbol{d}) = \boldsymbol{d}'\dot{\boldsymbol{T}}_0$, where $\dot{\boldsymbol{T}}_0 = \partial T_\omega/\partial\boldsymbol{\omega}$ evaluated at $\boldsymbol{\omega} = \boldsymbol{\omega}_0$.

Notice that the expressions of $\dot{\boldsymbol{T}}_0$ do not demand the estimator of $\boldsymbol{\theta}$ under the perturbed model ($\widehat{\boldsymbol{\theta}}_\omega$). Since $||\boldsymbol{d}|| = 1$, the maximum local slope is $||\dot{\boldsymbol{T}}_0||$ associated to the direction

$$\boldsymbol{d}_{\max} = -\frac{\dot{\boldsymbol{T}}_0}{||\dot{\boldsymbol{T}}_0||} \text{ (Wald)} \quad \text{or} \quad \boldsymbol{d}_{\max} = \frac{\dot{\boldsymbol{T}}_0}{||\dot{\boldsymbol{T}}_0||} \text{ (pivotal quantity)}. \tag{4.3}$$

Index plots of $\boldsymbol{d}_{\max}$ may point out those elements that under small perturbations exert high influence on $\widehat{\boldsymbol{\theta}}$ and $T$. Consequences of the perturbations can be assessed in a practical way by perturbing the model with $\boldsymbol{\omega} = \boldsymbol{\omega}_0 + a\boldsymbol{d}_{\max}$ and representing in graphs values of components of $\widehat{\boldsymbol{\theta}}$ and $T$ versus $a$.

## 4.4 Two Methods

Model formulation and inference are studied in Chap. 2. In the sequel, we present the elements of the matrix $\ddot{\boldsymbol{L}}$ of second derivatives of the profile log-likelihood function $l_{\mathrm{P}}(\boldsymbol{\theta})$ in (2.6) (common to all analyses involving the Wald statistic $W$), the matrix $\boldsymbol{\Delta}$ in (4.1) and the vector $\dot{\boldsymbol{T}}_0$ in (4.3) for different perturbation schemes proposed in de Castro et al. (2006b).

### *4.4.1 Matrix of Second Derivatives*

After some algebraic manipulations, we get from (2.6) the elements of $\ddot{\boldsymbol{L}}$, namely,

$$\frac{\partial^2 l_{\mathrm{P}}(\boldsymbol{\theta})}{\partial\alpha^2} = -\sum_{i=1}^{n} w_i,$$

$$\frac{\partial^2 l_{\mathrm{P}}(\boldsymbol{\theta})}{\partial\alpha\partial\beta} = \frac{\partial^2 l_{\mathrm{P}}(\boldsymbol{\theta})}{\partial\beta\partial\alpha} = -\sum_{i=1}^{n} \frac{(Y_i - \alpha - 2\beta\widehat{x}_i)\beta\kappa_i w_i + \widehat{x}_i}{\lambda_i}$$

and

$$\frac{\partial^2 l_{\mathrm{P}}(\boldsymbol{\theta})}{\partial\beta^2} = \sum_{i=1}^{n} \frac{(Y_i - \alpha - 2\beta\widehat{x}_i)^2\kappa_i w_i - \widehat{x}_i^2}{\lambda_i},$$

where $w_i$ and $\widehat{x}_i$ come from (2.8) and (2.4), respectively, $i = 1, \ldots, n$.

### 4.4.2 Perturbation of Sample Weights

In this scheme, to each summand in (2.6) is applied a weight resulting in

$$l_{\mathrm{P}}(\boldsymbol{\theta}, \boldsymbol{\omega}) = -\log(2\pi)\sum_{i=1}^{n}\omega_i - \frac{1}{2}\sum_{i=1}^{n}\omega_i \log(\kappa_i\lambda_i)$$
$$-\frac{1}{2}\sum_{i=1}^{n}\omega_i\left[\frac{X_i^2}{\kappa_i} + \frac{(Y_i-\alpha)^2}{\lambda_i} - \frac{1}{g_i}\left\{\frac{X_i}{\kappa_i} + \frac{(Y_i-\alpha)\beta}{\lambda_i}\right\}^2\right],$$

where $g_i$ is as in (2.5). This scheme generalizes the inclusion ($\omega_i = 1$) or the exclusion ($\omega_i = 0$) of a sample from the estimation of $\boldsymbol{\theta}$, so that this device enables us to learn about the importance of the samples. In this case, $n^* = n$ and $\boldsymbol{\omega}_0 = \mathbf{1}_n$.

The lines of the $\boldsymbol{\Delta}$ matrix in (4.1) are given by $\boldsymbol{\Delta}'_\alpha = (\boldsymbol{Y} - \alpha\mathbf{1}_n - \beta\boldsymbol{X}) \boxdot \boldsymbol{w}$ and $\boldsymbol{\Delta}'_\beta = \boldsymbol{\Delta}_\alpha \boxdot \widehat{\boldsymbol{x}}$, $\boldsymbol{X} = (X_1, \ldots, X_n)'$, "$\boxdot$" denotes the element-wise product, $\boldsymbol{w} = (w_1, \ldots, w_n)'$ and $\widehat{\boldsymbol{x}} = (\widehat{x}_1, \ldots, \widehat{x}_n)'$.

Analogously, from (2.18) the modified $\mathrm{Z}_0$ statistic takes the form $T_\omega = \mathrm{Z}_{0,\omega} = n^{-1/2}\sum_{i=1}^{n}\omega_i(Y_i - X_i)/(\lambda_i + \kappa_i)^{1/2}$. Hereafter, for the pivotal quantity statistic $\mathrm{Z}_0$, $\dot{\boldsymbol{T}}_0$ is denoted by $\dot{\boldsymbol{Z}}_0$. Therefore,

$$\dot{\boldsymbol{Z}}_0 = n^{-1/2}\left(\tfrac{Y_1 - X_1}{(\lambda_1+\kappa_1)^{1/2}}, \ldots, \tfrac{Y_n - X_n}{(\lambda_n+\kappa_n)^{1/2}}\right)'. \tag{4.4}$$

### 4.4.3 Perturbation of Measurements

In this section, the measurements taken from the old ("O") and new ("N") methods are modified through additive ("a") and multiplicative ("m") perturbation schemes. We can interpret additive and multiplicative disturbances of the measurements as absolute and relative changes of the data, respectively. Multiplicative disturbances are a natural expression of percentage variations. In this case, $n^* = n$. The no perturbation setting follows by taking $\boldsymbol{\omega}_0 = \mathbf{0}_n$ in the additive case and $\boldsymbol{\omega}_0 = \mathbf{1}_n$ in the multiplicative case.

### Old Method

The perturbations follow from (2.6) and (2.18) by putting

$$X_i(\omega_i) = \begin{cases} X_i + \omega_i, & \text{with additive perturbation,} \\ X_i\omega_i, & \text{with multiplicative perturbation,} \end{cases} \tag{4.5}$$

$i = 1, \ldots, n$. The required matrices and vectors are

$$\boldsymbol{\Delta}_{\mathrm{O}}^{\mathrm{a}} = \begin{bmatrix} \boldsymbol{\Delta}_{\mathrm{O}\alpha} \\ \boldsymbol{\Delta}_{\mathrm{O}\beta} \end{bmatrix}, \quad \boldsymbol{\Delta}_{\mathrm{O}}^{\mathrm{m}} = \boldsymbol{\Delta}_{\mathrm{O}}^{\mathrm{a}} \boxdot \begin{bmatrix} \boldsymbol{X}' \\ \boldsymbol{X}' \end{bmatrix}, \tag{4.6}$$

$$\dot{\boldsymbol{Z}}_{0,\mathrm{O}}^{\mathrm{a}} = -n^{-1/2}\left(\frac{1}{(\lambda_1+\kappa_1)^{1/2}}, \ldots, \frac{1}{(\lambda_n+\kappa_n)^{1/2}}\right)' \quad \text{and} \quad \dot{\boldsymbol{Z}}_{0,\mathrm{O}}^{\mathrm{m}} = \boldsymbol{X} \boxdot \dot{\boldsymbol{Z}}_{0,\mathrm{O}}^{\mathrm{a}}, \tag{4.7}$$

where $\boldsymbol{\Delta}_{\mathrm{O}\alpha}' = -\beta\boldsymbol{w}$ and $\boldsymbol{\Delta}_{\mathrm{O}\beta}' = \{-\beta\widehat{\boldsymbol{x}} + (\boldsymbol{Y} - \alpha\mathbf{1}_n - \beta\boldsymbol{X}) \boxdot \boldsymbol{\lambda} \boxdot \boldsymbol{w}\} \boxdot \boldsymbol{w}$, with $\boldsymbol{\lambda} = (\lambda_1, \ldots, \lambda_n)'$.

### New Method

Analogously to the perturbations in (4.5), the perturbations come from (2.6) and (2.18) by taking

$$Y_i(\omega_i) = \begin{cases} Y_i + \omega_i, & \text{with additive perturbation,} \\ Y_i\omega_i, & \text{with multiplicative perturbation,} \end{cases}$$

$i = 1, \ldots, n$. The matrices and vectors are

$$\boldsymbol{\Delta}_{\mathrm{N}}^{\mathrm{a}} = \begin{bmatrix} \boldsymbol{\Delta}_{\mathrm{N}\alpha} \\ \boldsymbol{\Delta}_{\mathrm{N}\beta} \end{bmatrix}, \quad \boldsymbol{\Delta}_{\mathrm{N}}^{\mathrm{m}} = \boldsymbol{\Delta}_{\mathrm{N}}^{\mathrm{a}} \boxdot \begin{bmatrix} \boldsymbol{Y}' \\ \boldsymbol{Y}' \end{bmatrix}, \tag{4.8}$$

$$\dot{\boldsymbol{Z}}_{0,\mathrm{N}}^{\mathrm{a}} = -\dot{\boldsymbol{Z}}_{0,\mathrm{O}}^{\mathrm{a}} = n^{-1/2}\left(\frac{1}{(\lambda_1+\kappa_1)^{1/2}}, \ldots, \frac{1}{(\lambda_n+\kappa_n)^{1/2}}\right)' \quad \text{and} \quad \dot{\boldsymbol{Z}}_{0,\mathrm{N}}^{\mathrm{m}} = \boldsymbol{Y} \boxdot \dot{\boldsymbol{Z}}_{0,\mathrm{N}}^{\mathrm{a}}, \tag{4.9}$$

where $\boldsymbol{\Delta}_{\mathrm{N}\alpha}' = \boldsymbol{w}$ and $\boldsymbol{\Delta}_{\mathrm{N}\beta}' = \{\widehat{\boldsymbol{x}} + \beta(\boldsymbol{Y} - \alpha\mathbf{1}_n - \beta\boldsymbol{X}) \boxdot \boldsymbol{\kappa} \boxdot \boldsymbol{w}\} \boxdot \boldsymbol{w}$, with $\boldsymbol{\kappa} = (\kappa_1, \ldots, \kappa_n)'$.

**Both Methods**

Another way of modifying the data is the joint perturbation of the measurements. The perturbations are obtained from (2.6) and (2.18 ) after replacing $(X_i, Y_i)'$ by

$$\begin{pmatrix} X_i(\omega_i) \\ Y_i(\omega_{i+n}) \end{pmatrix} = \begin{cases} \begin{pmatrix} X_i + \omega_i \\ Y_i + \omega_{i+n} \end{pmatrix}, & \text{with additive perturbation,} \\ \begin{pmatrix} X_i \omega_i \\ Y_i \omega_{i+n} \end{pmatrix}, & \text{with multiplicative perturbation,} \end{cases}$$

$i = 1, \ldots, n$. The vector of perturbations $\boldsymbol{\omega}$ is of dimension $2n$ $(n^* = 2n)$. In the additive case, $\boldsymbol{\omega}_0 = \mathbf{0}_{2n}$, whereas in the multiplicative case, $\boldsymbol{\omega}_0 = \mathbf{1}_{2n}$. The $\boldsymbol{\Delta}$ matrices in (4.1) result from combining the expressions just presented, that is, $\boldsymbol{\Delta}^{\mathrm{a}}_{\mathrm{ON}} = [\boldsymbol{\Delta}^{\mathrm{a}}_{\mathrm{O}}, \boldsymbol{\Delta}^{\mathrm{a}}_{\mathrm{N}}]$ and $\boldsymbol{\Delta}^{\mathrm{m}}_{\mathrm{ON}} = [\boldsymbol{\Delta}^{\mathrm{m}}_{\mathrm{O}}, \boldsymbol{\Delta}^{\mathrm{m}}_{\mathrm{N}}]$, where the four blocks are given in (4.6) and (4.8). In a similar fashion,

$$\dot{\boldsymbol{Z}}^{\mathrm{a}}_{0,\mathrm{ON}} = \begin{pmatrix} \dot{\boldsymbol{Z}}^{\mathrm{a}}_{\mathrm{O}} \\ \dot{\boldsymbol{Z}}^{\mathrm{a}}_{\mathrm{N}} \end{pmatrix} \quad \text{and} \quad \dot{\boldsymbol{Z}}^{\mathrm{m}}_{0,\mathrm{ON}} = \begin{pmatrix} \dot{\boldsymbol{Z}}^{\mathrm{m}}_{\mathrm{O}} \\ \dot{\boldsymbol{Z}}^{\mathrm{m}}_{\mathrm{N}} \end{pmatrix}, \tag{4.10}$$

with the required vectors found in (4.7) and (4.9).

The directions $\boldsymbol{d}_{\max}$ related to perturbations of the $\mathrm{Z}_0$ statistic in (2.18) are easier to analyze. Outstanding samples can be directly picked from the components of the vectors $\dot{\mathbf{Z}}_0$ in (4.4), (4.7), (4.9), and (4.10). Moreover, these schemes share a simple expression for the perturbed statistic $\mathrm{Z}_0(a)$, given by the linear function of $\mathrm{Z}_0(a) = \mathrm{Z}_0 + aS(\boldsymbol{d})$, $a \in \mathbb{R}$. In particular, if $\boldsymbol{d} = \boldsymbol{d}_{\max}$, then $\mathrm{Z}_0(a) = \mathrm{Z}_0 + a||\dot{\mathbf{Z}}_0||$ and if $\boldsymbol{d} = \boldsymbol{v}_{jn^*}$, $\mathrm{Z}_0(a) = \mathrm{Z}_0 + a\dot{\mathrm{Z}}_{0_j}$, $j = 1, \ldots, n^*$.

All the $\boldsymbol{\Delta}$ matrices are already evaluated at $\boldsymbol{\omega} = \boldsymbol{\omega}_0$ and must be computed with $\boldsymbol{\theta} = \widehat{\boldsymbol{\theta}}$.

## 4.5 Two or More Methods

Model formulation and inference are the subject of Chap. 3. Next, we provide the necessary matrices to implement some selected perturbation schemes studied in de Castro et al. (2007).

### 4.5.1 Perturbation of Sample Weights

In this scheme, to each summand in (3.6) is applied a weight, resulting in $l_P(\boldsymbol{\theta}, \boldsymbol{\omega}) = \sum_{i=1}^{n} \omega_i l_i(\boldsymbol{\theta})$, which generalizes the inclusion ($\omega_i = 1$) or the exclusion ($\omega_i = 0$) of an observation from the estimation of $\boldsymbol{\theta}$. In this case, $n^* = n$ and $\boldsymbol{\omega}_0 = \mathbf{1}_n$.

The blocks of the $\boldsymbol{\Delta}$ matrix in (4.1) are given by $\boldsymbol{\Delta}_\alpha = [\boldsymbol{D}^{-1}(\boldsymbol{\lambda}_1)(\boldsymbol{Y}_{1\bullet} - \boldsymbol{\alpha} - \boldsymbol{\beta}\widehat{x}_1), \ldots, \boldsymbol{D}^{-1}(\boldsymbol{\lambda}_n)(\boldsymbol{Y}_{n\bullet} - \boldsymbol{\alpha} - \boldsymbol{\beta}\widehat{x}_n)]$ and $\boldsymbol{\Delta}_\beta = [\boldsymbol{D}^{-1}(\boldsymbol{\lambda}_1)(\boldsymbol{Y}_{1\bullet} - \boldsymbol{\alpha} - \boldsymbol{\beta}\widehat{x}_1)\widehat{x}_1, \ldots, \boldsymbol{D}^{-1}(\boldsymbol{\lambda}_n)(\boldsymbol{Y}_{n\bullet} - \boldsymbol{\alpha} - \boldsymbol{\beta}\widehat{x}_n)\widehat{x}_n]$, where $\widehat{x}_i$ is as in (3.4) and $\boldsymbol{Y}_{i\bullet} = (Y_{i1}, \ldots, Y_{ir})'$, $i = 1, \ldots, n$.

### 4.5.2 Perturbation of Measurements

In this section, the measurements ($X$ and $\boldsymbol{Y}$) are modified through additive ("a") and multiplicative ("m") perturbation schemes. In this case, $n^* = n$. We turn back to the working model by taking $\boldsymbol{\omega}_0 = \mathbf{0}$ in the additive case and $\boldsymbol{\omega}_0 = \mathbf{1}_n$ in the multiplicative case.

**Old Method**

The perturbed models follow from (3.6) by putting

$$X_i(\omega_i) = \begin{cases} X_i + \omega_i, & \text{with additive perturbation,} \\ X_i\omega_i, & \text{with multiplicative perturbation,} \end{cases}$$

$i = 1, \ldots, n$. The required matrices are

$$\boldsymbol{\Delta}^{\mathrm{a}} = \sum_{i=1}^{n} \frac{1}{g_i \kappa_i} \begin{pmatrix} -\boldsymbol{D}^{-1}(\boldsymbol{\lambda}_i)\boldsymbol{\beta} \\ \boldsymbol{D}^{-1}(\boldsymbol{\lambda}_i)(\boldsymbol{Y}_{i\bullet} - \boldsymbol{\alpha} - 2\widehat{x}_i\boldsymbol{\beta}) \end{pmatrix} \boldsymbol{v}'_{in} \quad \text{and} \quad \boldsymbol{\Delta}^{\mathrm{m}} = \boldsymbol{\Delta}^{\mathrm{a}} \boxdot (\mathbf{1}_{2r}\boldsymbol{X}'),$$

where $g_i$ comes from (3.5).

**New Methods**

We perturb the measurements of the $j$-th new method $\boldsymbol{Y}_{\bullet j} = (Y_{1j}, \ldots, Y_{nj})'$ leading to

$$Y_{ij}(\omega_i) = \begin{cases} Y_{ij} + \omega_i, & \text{with additive perturbation,} \\ Y_{ij}\omega_i, & \text{with multiplicative perturbation,} \end{cases} \tag{4.11}$$

$i = 1, \ldots, n$. Replacing $Y_{ij}(\omega_i)$ in (3.6), we obtain the perturbed log-likelihood function and arrive at

$$\boldsymbol{\Delta}^{\mathrm{a}} = \sum_{i=1}^{n} \frac{1}{\lambda_{ij}} \begin{pmatrix} \boldsymbol{v}_{jr} - g_i^{-1}\beta_j \boldsymbol{D}^{-1}(\boldsymbol{\lambda}_i)\boldsymbol{\beta} \\ g_i^{-1}\beta_j \boldsymbol{D}^{-1}(\boldsymbol{\lambda}_i)(\boldsymbol{Y}_{i\bullet} - \boldsymbol{\alpha} - 2\widehat{x}_i\boldsymbol{\beta}) + \widehat{x}_i \boldsymbol{v}_{jr} \end{pmatrix} \boldsymbol{v}_{in}'$$

and $\boldsymbol{\Delta}^{\mathrm{m}} = \boldsymbol{\Delta}^{\mathrm{a}} \boxdot (\mathbf{1}_{2r}\boldsymbol{Y}_{\bullet j}')$, for $j \in \{1, \ldots, r\}$.

Similar to Sect. 4.4.3, measurements taken with the old method and one of the new methods might also be perturbed. Moreover, perturbations of the measurements from all the $r$ new methods or from the $r + 1$ methods give rise to different perturbation schemes.

With respect to the pivotal quantity statistic in (3.20), the generalization of the vectors $\dot{\boldsymbol{Z}}_0$ in (4.4), (4.7), (4.9), and (4.10) is simple. For example, the vector in (4.4) has components $(nr)^{-1/2}(Y_{ij} - X_i)/(\lambda_{ij} + \kappa_i)$, for $i = 1, \ldots, n$ and $j = 1, \ldots, r$.

All the $\boldsymbol{\Delta}$ matrices in Sects. 4.4 and 4.5 are already evaluated at $\boldsymbol{\omega} = \boldsymbol{\omega}_0$ and must be evaluated at $\boldsymbol{\theta} = \widehat{\boldsymbol{\theta}}$. It is worth remarking that the expressions for the influence assessment in Sects. 4.3–4.5 are suitable for implementation in a matrix language.

# Chapter 5
# Data Analysis

## 5.1 Introduction

Two data sets comprising two and three measurement methods are analyzed. The tolerance for the convergence criterion in (3.11) is set at $10^{-5}$. All the computations were carried out in the R language (R Core Team 2019). Numerical derivatives in Sect. 4.3 were computed using the numDeriv package (Gilbert and Varadhan 2016). In our analyses, $p$-value $< 0.05$ was adopted to determine statistical significance. R scripts for the data analyses in this chapter are listed in Appendix B.

## 5.2 Two Methods

Thirty pairs of determinations of the arsenate ion in natural river water by two methods described as (1) continuous selective reduction and atomic absorption spectrometry (the "old" method) and (2) non-selective reduction, cold trapping, and atomic emission spectrometry (the "new" method). Each determination is the mean of three replicates, from which the variances of the errors are also obtained. The data are presented in Ripley and Thompson (1987, Table 1). The concentrations vary from 0 to 20 μg/l. One sample is such that determinations by both methods are 0.00. After discarding this sample, maximum likelihood estimates (standard errors) are $\widehat{\alpha} = 0.347$ (0.0854) and $\widehat{\beta} = 0.843$ (0.0816), while corrected score estimates (standard errors) are $\widetilde{\alpha} = 0.240$ (0.139) and $\widetilde{\beta} = 0.828$ (0.0565).

Determinations, standard deviations of the measurement errors, and fitted regression lines are shown in Fig. 5.1. From the model checking tool in Sect. 4.2, the quantile-quantile (QQ) plot and simulated envelope in Fig. 5.2 with $M = 100$ simulated samples suggest that the model in Sect. 2.2 yields a good fit.



H. Bolfarine et al., *Regression Models for the Comparison of Measurement Methods*, SpringerBriefs in Statistics, https://doi.org/10.1007/978-3-030-57935-7_5

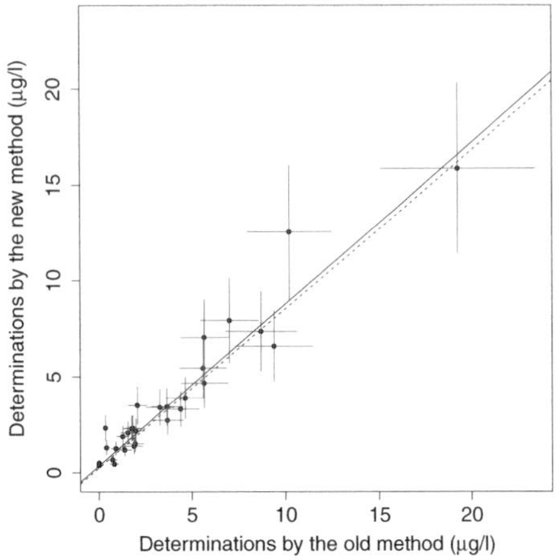


**Fig. 5.1** Determinations, standard deviations of the measurement errors (blue plus symbol), and fitted regression lines (solid: maximum likelihood and dashed: corrected score)

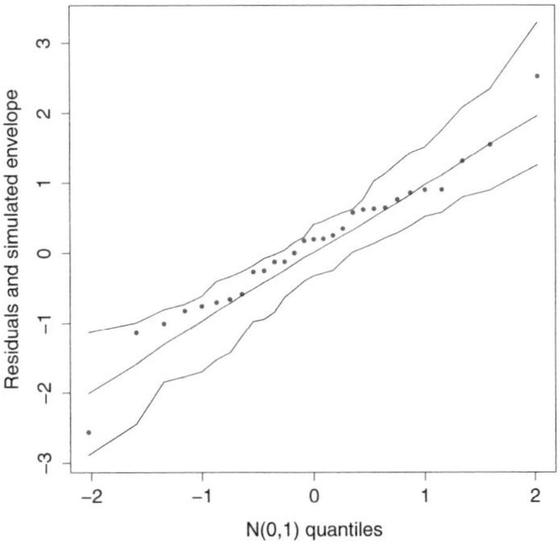


**Fig. 5.2** QQ plot of the residuals and simulated envelope

The Wald statistics in (2.12) and (2.15) lead us to reject $H_0$ in (2.11) ($W = 16.35$, $p = 0.00028$ and $W_{CS} = 9.451$, $p = 0.00887$), whereas according to the exact test in Sect. 2.6, $H_0$ is not rejected, as can be seen in Fig. 5.3 ($Z_0 = 1.718$, $p = 0.08576$). Results from simulation studies in de Castro et al. (2005) suggest that departures from $H_0$ in (2.11) like the ones in the maximum likelihood estimates are detected with greater power by the Wald statistic in (2.12). In what follows, the results refer to the Wald statistic in (2.12).

The level of agreement between the measurement methods is assessed through the probability of agreement plot in Sect. 2.7. Figure 5.4 was built with $x_i^* = \widehat{x}_i$ and the corresponding error variances $\lambda_i$ and $\kappa_i$, $i = 1, \ldots, 29$. The maximal difference $\epsilon = 0.54\mu g/l$ in (2.19) corresponds to the median of $|Y_i - X_i|$, $i = 1, \ldots, 29$. For most of the samples, the estimate of the probability of agreement is smaller than 0.95, which means low agreement between the methods, and decreases with

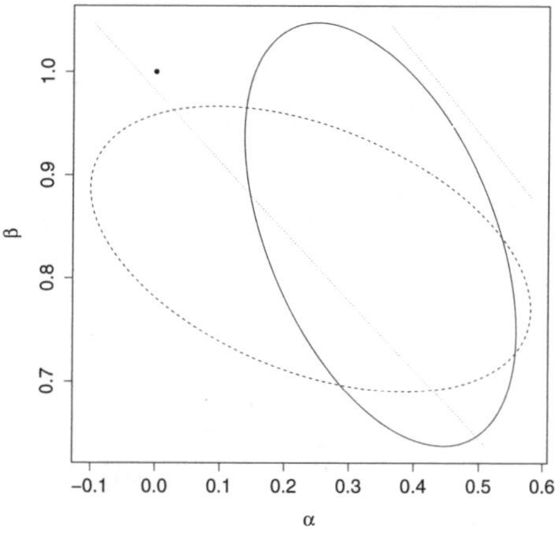


**Fig. 5.3** Ninety five percentage confidence regions for $\alpha$ and $\beta$ and the point indicating absence of biases (solid: maximum likelihood, dashed: corrected score, and dotted: pivotal quantity)

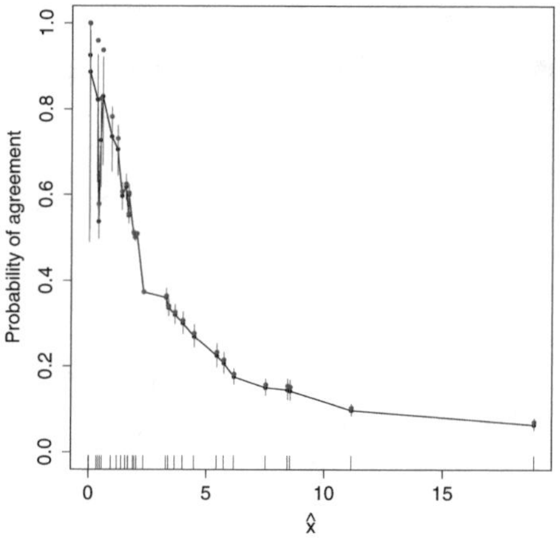


**Fig. 5.4** Probability of agreement under $H_0$ (blue filled circle), probability of agreement plot, and 95% confidence interval (vertical red line) with $\epsilon = 0.54\,\mu$g/l

the estimate of the true determination. In general, there is low agreement between the methods.

In the sequel, we apply local influence methods described in Sects. 4.3 and 4.4. Perturbing the sample weights, Fig. 5.5a highlights samples 19 and 22. Since the signs of the 19th and 22nd components of $\boldsymbol{d}_{\text{max}}$ are opposite ($d_{\text{max},19} = -0.7505$ and $d_{\text{max},22} = 0.4559$), the greatest changes in the Wald statistic $W$ in (2.12) are obtained by perturbing the weights of these samples in opposite directions starting at $\boldsymbol{\omega}_0 = \mathbf{1}_{29}$. In Fig. 5.5b the direction $\boldsymbol{d}_{\text{max}}$ is quite different. We assess the effect of deleting samples one a time on the values of $W$ and $|\mathrm{Z}_0|$. Complementary to Figs. 5.5 and 5.6 displays the statistics $W$ and $|\mathrm{Z}_0|$ computed after deleting one sample a time, denoted by $W_{-i}$ and $|\mathrm{Z}_{0,-i}|$. Only the removal of sample 19 would lead to a change in the decision about the absence of bias, but $|\mathrm{Z}_{0,-19}| = 2.020$ is borderline ($p = 0.04334$).

With respect to additive perturbations determinations from the old method, index plots of $|\boldsymbol{d}_{\text{max}}|$ in Fig. 5.7a, c reveal similarity between the Wald and pivotal

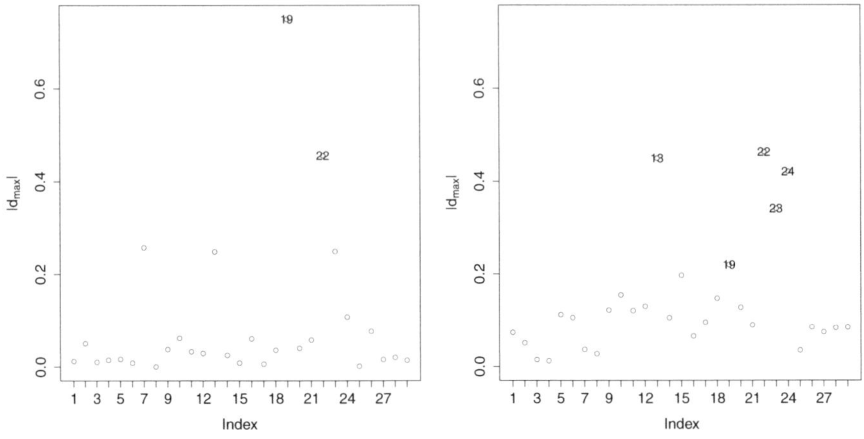


(a) (b)

**Fig. 5.5** Index plots of $|\boldsymbol{d}_{\max}|$ for perturbation of sample weights: (**a**) Wald and (**b**) pivotal quantity

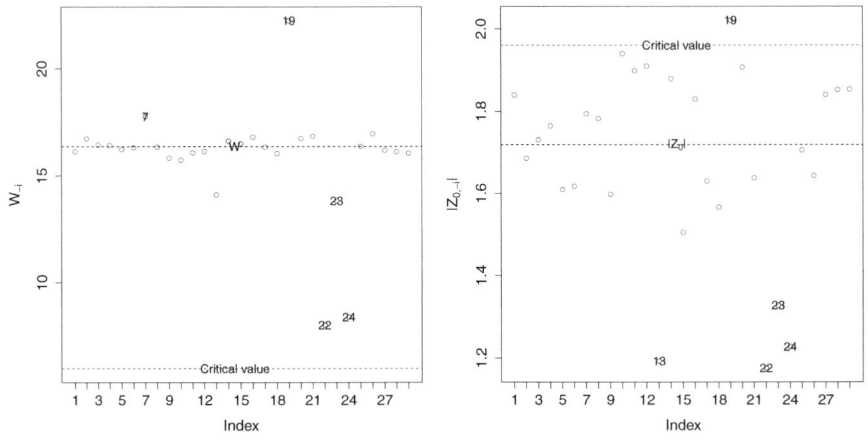


(a) (b)

**Fig. 5.6** Test statistics after removing one sample a time: (**a**) Wald and (**b**) pivotal quantity (dashed black lines represent the statistics computed with the full data set) statistics under additive perturbations. Samples 22 and 24 ($X_{22} = X_{24} = 0.00$) deserve attention. In Fig. 5.7b, samples 7 and 19 are outstanding. The direction $\boldsymbol{d}_{\max}$ in Fig. 5.7b does not contain a small subset of outstanding samples, so that there is no sample or small subset of samples potentially influent on $Z_0$. Compared to Fig. 5.7, disturbances of determinations from the new method have directions $\boldsymbol{d}_{\max}$ in Fig. 5.8 with an analogous pattern. This is not surprising, because determinations and standard deviations from the old and new methods are not so different (see Fig. 5.1). As expected, results of perturbations of determinations from both methods are some kind of combination of the two previous results (Figs. 5.7 and 5.8), and we omit the results.

We further examine the effects of departures of $\boldsymbol{\omega}_0 = \mathbf{0}_{29}$ along the direction $\boldsymbol{d}_{\max}$ in Fig. 5.7a. The direction $\boldsymbol{d}_{\max,22,24}$ is obtained by replacing the components of $\boldsymbol{d}_{\max}$ by 0, except samples 22 and 24, and renormalizing it so that $||\boldsymbol{d}_{\max,22,24}|| = 1$. To draw Fig. 5.9, for each selected value of $a$ we calculate $X_i(\omega_i) = X_i + a\boldsymbol{d}_{\max,i}$, $i = 1, \ldots, 29$, $\alpha$ and $\beta$ are estimated leading to $W(a)$. Solid and dashed curves in Fig. 5.9 are similar, suggesting that the changes in $W$ are dominated by samples 22 and 24. Sample 24 alone in direction $-\boldsymbol{v}_{24,29}$ (we take the negative sign because $d_{\max,24} < 0$) does not produce the same effect. Perturbing along the direction $\boldsymbol{d}_{\max,22,24}$, if $X_{22}$ and $X_{24}$ were 0.17 and 0.24 μg/l instead of 0.00, which corresponds to $a = -0.3$ in Fig. 5.9, the value of $W$ would be reduced from 16.35 to 5.056 and the hypothesis $H_0$ in (2.11) would not be rejected ($p = 0.07982$). This brings to light the importance of few samples on a decision.

## 5.3 Three Methods

Now we give an illustrative example of the methodology developed in Sects. 4.3 and 4.5. Lao et al. (1996, Table 5) present recoveries of polycyclic aromatic hydrocarbons (PAH) from ambient-air polyurethane foam (PUF) samples (in %) extracted by three methods, namely, microwave-assisted process (MAP), Soxhlet extraction, and mechanical extraction. This data comprises 22 samples. In our analysis, MAP is the old method. Data, standard deviations, as well as the maximum likelihood regression lines are shown in Fig. 5.10.

The biases $(\alpha_1, \beta_1)'$ and $(\alpha_2, \beta_2)'$ correspond to the Soxhlet and mechanical extraction methods, respectively. Maximum likelihood estimates (and standard errors) are $\widehat{\alpha}_1 = 22.20$ (10.82), $\widehat{\beta}_1 = 0.758$ (0.127), $\widehat{\alpha}_2 = -18.16$ (12.74), and $\widehat{\beta}_2 = 0.947$ (0.148). Applying the model checking tool in Sect. 4.2, the QQ plot and simulated envelope in Fig. 5.11 with $M = 100$ simulated samples suggest that the model in Sect. 3.2 yields a good fit.

The Soxhlet and mechanical extraction methods can be compared to the MAP method by testing $H_0 : (\boldsymbol{\alpha}', \boldsymbol{\beta}')' = (\mathbf{0}', \mathbf{1}_2')'$, as in (3.16). This hypothesis is rejected, for the statistic $W$ in (2.12) takes the value 541.6 ($p < 0.00001$). The hypothesis in (3.17) corresponds to separate tests for the Soxhlet and mechanical extraction methods, which is tested with the statistic in (3.18). Applying the test, we obtain

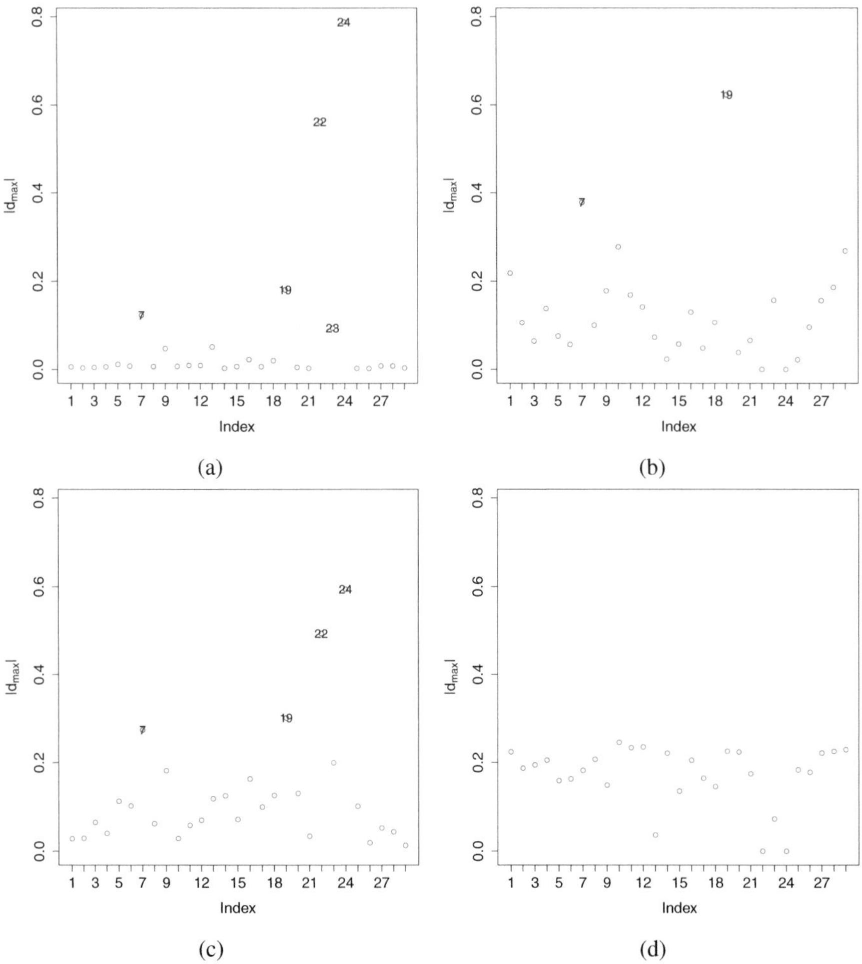


**Fig. 5.7** Index plots of $|\boldsymbol{d}_{\max}|$ for perturbation of determinations from the old method: (**a**) Wald, additive, (**b**) Wald, multiplicative, (**c**) pivotal quantity, additive, and (**d**) pivotal quantity, multiplicative

$W_1 = 5.009$ ($p = 0.08173$) and $W_2 = 207.0$ ($p < 0.00001$) so that there is no significant biases in the recoveries from the Soxhlet method.

The probability of agreement plot in Sect. 3.6 is used to assess the level of agreement between each new measurement method (Soxhlet extraction and mechanical extraction) and the old method. Figure 5.12 was built with $x_i^* = \widehat{x}_i$ and the corresponding error variances $\lambda_{ij}$ and $\kappa_i$, for $i = 1, \ldots, 22$ and $j = 1, 2$. The maximal differences $\epsilon = 4.2\%$ and $\epsilon = 15.95\%$ in (3.21) correspond to the medians of $|Y_{i1} - X_i|$ and $|Y_{i2} - X_i|$, $i = 1, \ldots, 22$, respectively. For most of the

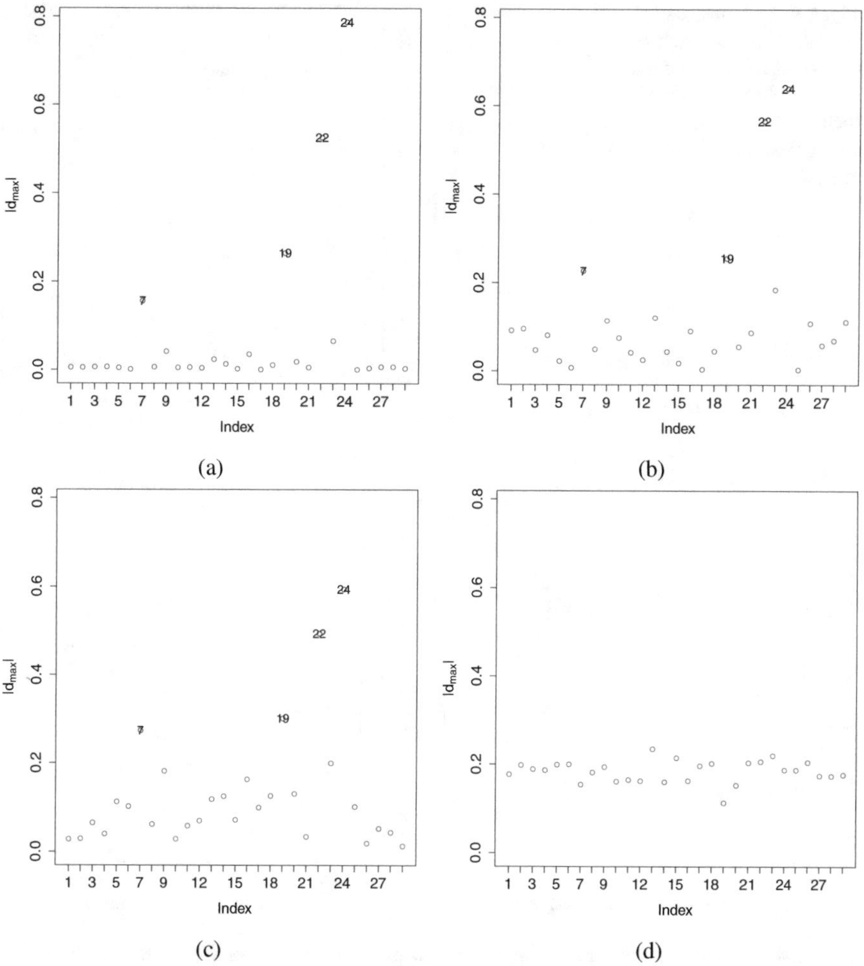


**Fig. 5.8** Index plots of $|d_{\max}|$ for perturbation of determinations from the new method: (**a**) Wald, additive, (**b**) Wald, multiplicative, (**c**) pivotal quantity, additive, and (**d**) pivotal quantity, multiplicative

samples, the estimates of the probability of agreement are smaller than 0.95, which means low agreement between the methods.

In Fig. 5.12 we see that, as pointed out by Choudhary and Nagaraja (2017) and Stevens et al. (2017), if a method is unbiased, absence of biases does not imply that the probability of agreement is necessarily high. And vice versa, if a method is biased, presence of bias does not imply that the probability of agreement is low. As mentioned above, measures of agreement (Choudhary and Nagaraja 2017; Stevens et al. 2017) strongly depend on the variances of the measurement errors, that is,

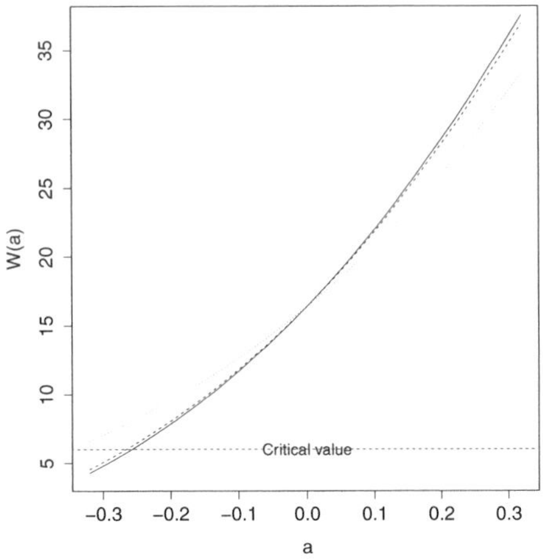


**Fig. 5.9** Wald test statistic versus $a$ along selected directions for perturbation of determinations from the old method (solid: $\boldsymbol{d}_{\max}$, dashed: $\boldsymbol{d}_{\max,22,24}$, and dotted: $\boldsymbol{v}_{24,29}$)

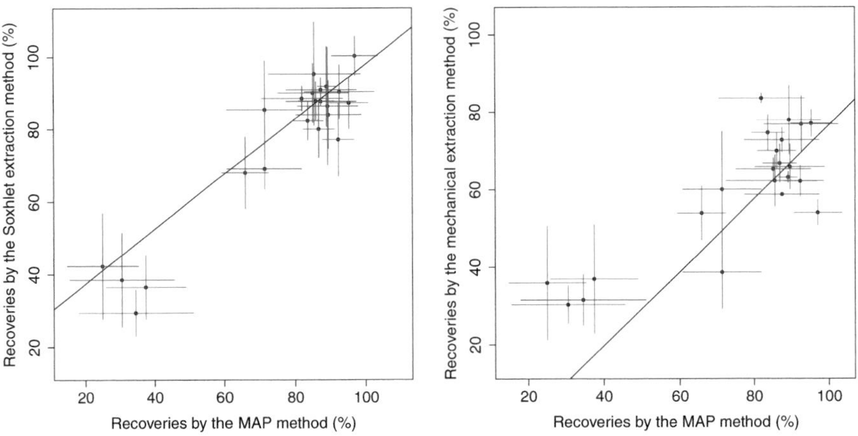


**Fig. 5.10** Recoveries, standard deviations of the measurement errors (blue plus symbol), and maximum likelihood fitted regression lines

two methods can be identical ($\alpha = 0$, $\beta = 1$, and $\lambda = \kappa$) and have a low level of agreement, and vice versa.

Now we apply local influence methods described in Sects. 4.3 and 4.5. Our interest is in the Wald statistic $W$ in (2.12). Perturbing the sample weights, Fig. 5.13a highlights samples 2, 11, and 21. The effect of deleting samples one a time on the values of $W$ can be seen in Fig. 5.13b. We see that the deletion of no particular sample induces a change in the decision about the absence of bias.

With respect to perturbations of recoveries from the old method (MAP method), index plots of $|\boldsymbol{d}_{\max}|$ in Fig. 5.14 reveal similarity between additive and multiplicative perturbations. Sample 13 stands out as potentially influent, noticing that this sample has the greatest value of $X_i/\kappa_i$, for $i = 1, \ldots, 22$.

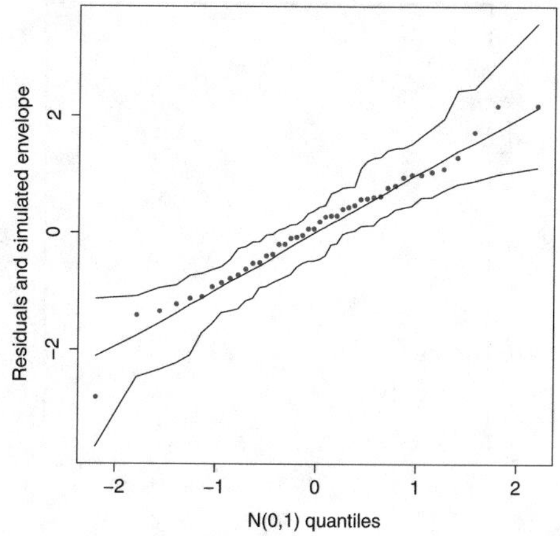


**Fig. 5.11** QQ plot of the residuals and simulated envelope

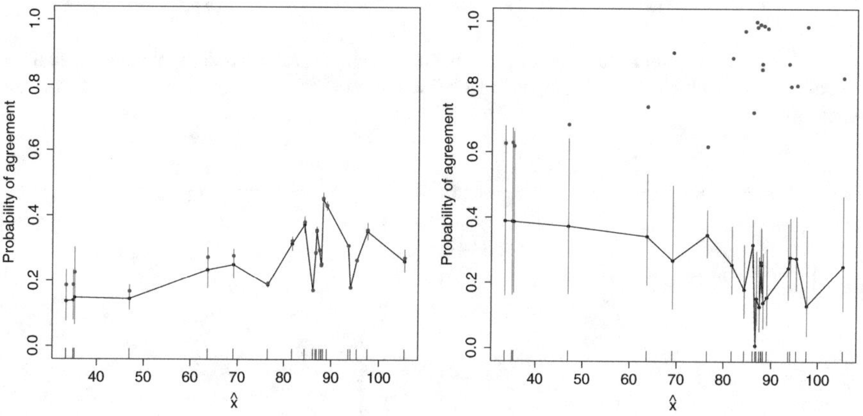


(a) (b)

**Fig. 5.12** Probability of agreement under $H_0$ (blue filled circle), probability of agreement plot, and 95% confidence interval (vertical red line): (**a**) Soxhlet extraction method with $\epsilon = 4.2\%$ and (**b**) mechanical extraction method with $\epsilon = 15.95\%$

Looking at perturbations of recoveries from the new methods, index plots of $|\boldsymbol{d}_{\max}|$ in Fig. 5.15 reveal similarity between additive and multiplicative perturbations. Samples 11 and 21 are far way and are potentially influent. Notice that these samples have the greatest value of $Y_{ij}/\lambda_{ij}$, for $i = 1, \ldots, 22$ and both new methods. However, changes in the values of the Wald statistic after perturbations along the directions $\boldsymbol{d}_{\max}$ and $\boldsymbol{d}_{\max,11,21}$ over a wide range of values of $a$ do not cause a change in the decision of rejecting the hypothesis in (3.16). Because of that, we omit the results.

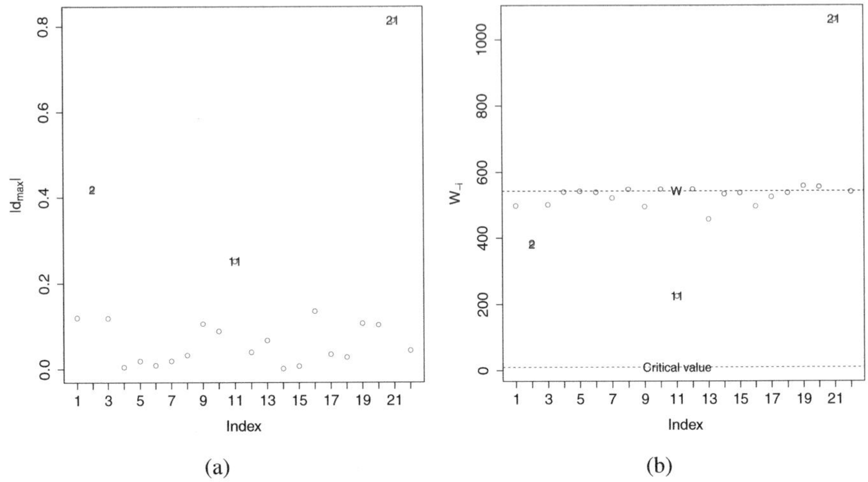


**Fig. 5.13** (**a**) Index plot of $|\boldsymbol{d}_{\max}|$ for perturbation of sample weights and (**b**) Wald statistic after removing one sample a time (the dashed black line represents the statistic computed with the full data set)

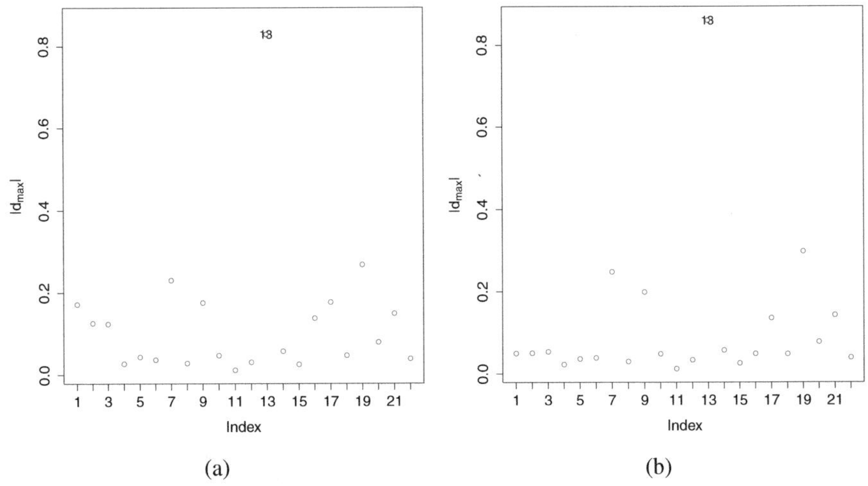


**Fig. 5.14** Index plots of $|\boldsymbol{d}_{\max}|$ for perturbation of recoveries from the old method: (**a**) additive and (**b**) multiplicative

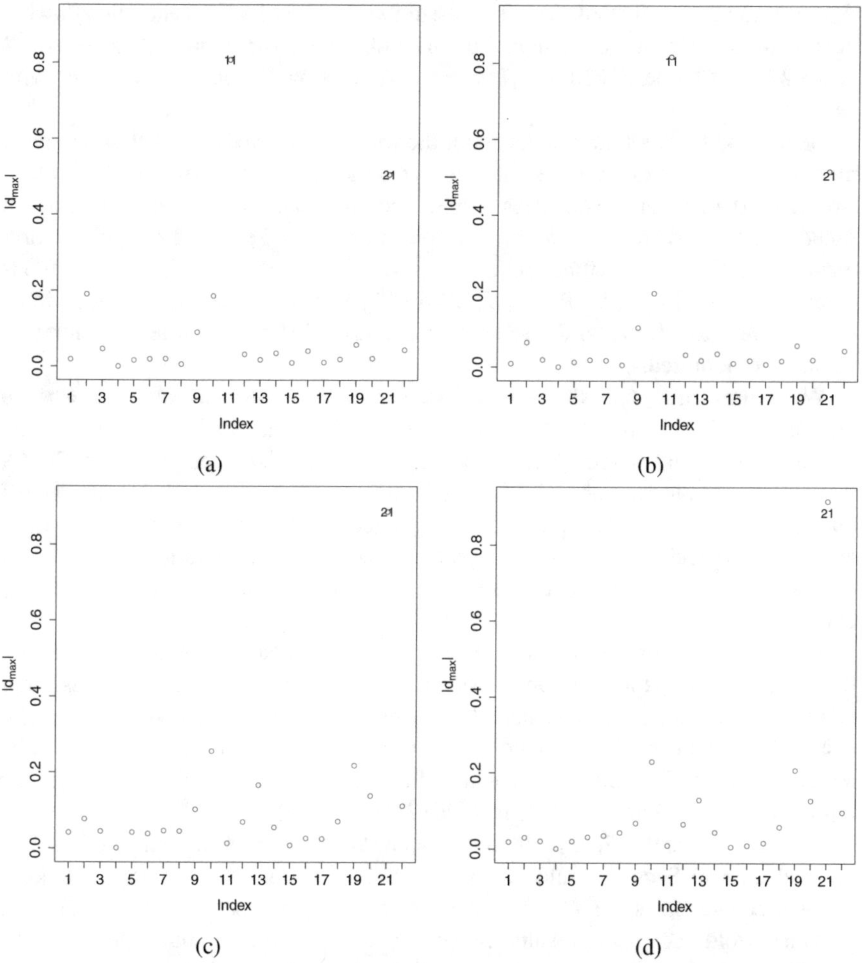


**Fig. 5.15** Index plots of $|\boldsymbol{d}_{\max}|$ for perturbation of recoveries from the new methods: (**a**) additive, Soxhlet method, (**b**) multiplicative, Soxhlet method, (**c**) additive, mechanical method, and (**d**) multiplicative, mechanical method

## 5.4 Conclusion

The local influence techniques provide to the practitioner numerical and graphical tools that are useful for the analysis. Influence analysis is recognized to be an essential part of the analysis, but it has received only scarce attention in the literature. Based on these techniques, it is possible to identify potential influential samples and how to assess the effects of perturbations on test statistics.

Additional perturbation schemes might be investigated. Until now, variances of the measurement errors were assumed to be the known constants, which is a key aspect of the model formulation. This assumption might be weakened by allowing disturbances from this condition. For instance, the error variances in Sects. 2.2 and 3.2 might be perturbed as $\kappa_i(\omega_i) = \kappa_i/\omega_i$, $\omega_i > 0$, for $i = 1, \ldots, n$, with $\boldsymbol{\omega}_0 = \mathbf{1}_n$.

Depending on the assumptions about the true (unobserved) concentration value, measurement error models are classified as either functional, structural, or ultra-structural (Cheng and Van Ness 1999, Section 1.1). The distinction between functional and structural *modeling* (Carroll et al. 2006, Section 2.1) is also worth mentioning. We embrace functional models, which are commonly found in methods comparison problems (del Río et al. 2001; Cheng and Riu 2006). The functional model is less restrictive in the sense that a distribution for the true concentration value is not required.

We assume throughout the book that the variances of the measurement errors are known. Notwithstanding this limitation, the models in this book have proven to be useful in many real-world applications. As pointed out by Cheng and Riu (2006), information on the variances can be obtained from (1) similar experiments carried out in the past, (2) from replicated data gauged in an independent similar study, and (3) from precision studies, among others. Replicated observations, if available, enable us to propose a model comprising unknown error variances; see, e.g., Dolby et al. (1987).

For the sake of simplicity and with the literature on methods comparison in mind, the normal model is adopted in the book. A graphical tool for model assessment is provided in Sect. 4.2. The models can be extended to the class of elliptical distributions (Fang et al. 1990), which includes the normal distribution. Specifically, de Castro and Galea (2010) deal with a Student $t$ model (see also the model for replicated data in Galea and de Castro 2017).

In Sects. 2.8 and 3.7, the simulated scenarios mimic some characteristics of the real data in Sects. 5.2 and 5.3. Under different scenarios, simulation studies summarized in Sects. 2.8 and 3.7 indicate that the proposed estimators and test statistics yield acceptable results. Since it is not possible to enunciate a general theoretical result regarding the merits of the estimators and the tests (covering sample size and extent of the error variances heteroscedasticity), as a guidance to the practitioner we suggest a choice based on simulations similar to the conditions of the problem at hand.

# Appendix A
# Miscellaneous Results

## A.1 Corrected Score Approach

In a sample of size $n$, consider an $n \times 1$ response vector $\boldsymbol{Y} = (Y_1, \ldots, Y_n)'$, an $n \times 1$ covariate vector $\boldsymbol{x} = (x_1, \ldots, x_n)'$, and a $p \times 1$ parameter vector $\boldsymbol{\theta}$. Let $l(\boldsymbol{\theta}; \boldsymbol{x}, \boldsymbol{Y}) = \sum_{i=1}^{n} l_i(\boldsymbol{\theta}; x_i, Y_i)$ be the log-likelihood function of $\boldsymbol{\theta}$ for observed data $\boldsymbol{x}$ and $\boldsymbol{Y}$. Under appropriate regularity conditions (Lehmann 1998), the score function is given by

$$\boldsymbol{U} = \boldsymbol{U}(\boldsymbol{\theta}; \boldsymbol{x}, \boldsymbol{Y}) = \frac{\partial}{\partial \boldsymbol{\theta}} l(\boldsymbol{\theta}; \boldsymbol{x}, \boldsymbol{Y}). \tag{A.1}$$

Moreover, the maximum likelihood estimator $\widehat{\boldsymbol{\theta}}$ of $\boldsymbol{\theta}$ follows by maximizing the likelihood function and it can be obtained by solving the estimating equation $\boldsymbol{U} = \boldsymbol{0}_p$ that follows from (A.1). Under those conditions, $\widehat{\boldsymbol{\theta}}$ is consistent and asymptotically normal, with asymptotic covariance matrix given by $\{\mathrm{E}(-\partial \boldsymbol{U}/\partial \boldsymbol{\theta})\}^{-1}$. Now, since $\boldsymbol{x}$ cannot be observed, as in the situation where it is measured with error (see the models in Chaps. 2 and 3), the above approach is not useful. For example, in the case of additive measurement errors $u_i$, where $X_i = x_i + u_i$, we can replace $x_i$ for $X_i$ in (A.1), $i = 1, \ldots, n$, that is, considering the naive score estimating equation $\boldsymbol{U}(\boldsymbol{\theta}; \boldsymbol{X}, \boldsymbol{Y}) = \boldsymbol{0}_p$, where $\boldsymbol{X} = (X_1, \ldots, X_n)'$, leading to the naive approach which does not yield a consistent estimator of $\boldsymbol{\theta}$ (Carroll et al. 2006, Chapter 3).

To overcome such difficulties, in Stefanski (1989) and Nakamura (1990) is proposed the corrected likelihood approach, which consists in obtaining a corrected log-likelihood function $l^*(\boldsymbol{\theta}; \boldsymbol{X}, \boldsymbol{Y}) = \sum_{i=1}^{n} l_i^*(\boldsymbol{\theta}; X_i, Y_i)$ such that $\mathrm{E}\{l^*(\boldsymbol{\theta}; \boldsymbol{X}, \boldsymbol{Y})|\boldsymbol{x}, \boldsymbol{Y}\} = l(\boldsymbol{\theta}; \boldsymbol{X}, \boldsymbol{Y})$, which will lead to a consistent estimator of $\boldsymbol{\theta}$ obtained by using the corrected score function given by $\boldsymbol{U}^* = \boldsymbol{U}^*(\boldsymbol{\theta}; \boldsymbol{X}, \boldsymbol{Y}) = \sum_{i=1}^{n} \boldsymbol{U}_i^*$, where $\boldsymbol{U}_i^* = \boldsymbol{U}_i^*(\boldsymbol{\theta}; X_i, Y_i) = \partial l_i^*(\boldsymbol{\theta}; X_i, Y_i)/\partial \boldsymbol{\theta}$. Thus, the corrected



H. Bolfarine et al., *Regression Models for the Comparison of Measurement Methods*, SpringerBriefs in Statistics, https://doi.org/10.1007/978-3-030-57935-7

score estimator of $\boldsymbol{\theta}$, denoted by $\widetilde{\boldsymbol{\theta}}$, follows by solving the equation $\boldsymbol{U}^* = \mathbf{0}_p$. Asymptotic properties of the estimator $\widetilde{\boldsymbol{\theta}}$ are discussed in the sequel.

## A.2 Unbiased Estimating Equations

The estimation methods in Chaps. 2 and 3 can be formulated under the unbiased estimating equations approach (see, e.g., Carroll et al. 2006, Section A.6). Maximum likelihood ($\widehat{\boldsymbol{\theta}}$) and corrected score estimators ($\widetilde{\boldsymbol{\theta}}$) are solutions of the unbiased estimating equations $\sum_{i=1}^n \boldsymbol{\psi}_i = \mathbf{0}$, where $\boldsymbol{\psi}_i = \boldsymbol{\psi}_i(\boldsymbol{\theta}, \boldsymbol{Z}_i)$, with $\boldsymbol{\psi}_i = \boldsymbol{q}_i = \partial h_i(\boldsymbol{\theta})/\partial\boldsymbol{\theta}$ and $\boldsymbol{\psi}_i(\boldsymbol{\theta}, \boldsymbol{Z}_i) = \boldsymbol{U}_i^* = \partial l_i^*(\boldsymbol{\theta})/\partial\boldsymbol{\theta}$, respectively (see Sects. 2.3, 2.5, 3.3, and 3.4).

Under suitable regularity conditions, maximum likelihood and corrected score estimators are consistent and asymptotically normally distributed with mean vector $\boldsymbol{\theta}^0$ and covariance matrix $n^{-1}\boldsymbol{\Omega} = \boldsymbol{A}^{-1}\boldsymbol{V}(\boldsymbol{A}^{-1})'$, where $\boldsymbol{A} = n^{-1}\sum_{i=1}^n \mathrm{E}(-\partial\boldsymbol{\psi}_i/\partial\boldsymbol{\theta})$ and $\boldsymbol{V} = n^{-1}\sum_{i=1}^n \mathrm{E}(\boldsymbol{\psi}_i\boldsymbol{\psi}_i') = n^{-1}\sum_{i=1}^n \mathrm{cov}(\boldsymbol{\psi}_i)$. The expectations are taken with respect to the true parameter values $\boldsymbol{\theta}^0$ and $x_1^0, \ldots, x_n^0$. With respect to the regularity conditions, the normality assumption ensures that all the required expectations are finite. Two key assumptions are (1) $0 < \liminf_{n\to\infty}\sum_{i=1}^n (x_i - \overline{x})^2/n \le \limsup_{n\to\infty}\sum_{i=1}^n (x_i - \overline{x})^2/n < \infty$, where $\overline{x} = \sum_{i=1}^n x_i/n$, and (2) the existence of $\rho > 0$ such that $\lim_{n\to\infty}\sum_{i=1}^n |x_i|^{2+\rho}/n^{1+\rho/2} = 0$. The first condition requires that the true unobservable $x$ should be neither too spread nor concentrated when $n \to \infty$. The second assumption allows an application of the Lyapunov's central limit theorem in order to obtain the asymptotic distribution of the estimator of $\boldsymbol{\theta}$.

We can mention two ways to estimate the covariance matrix $n^{-1}\boldsymbol{\Omega}$. For the maximum likelihood method in Sects. 2.3.1 and 3.3.1, we adopt the so-called model-based expectation method. With this method, we replace $\boldsymbol{\theta}$, $x_i$ and $x_i^2$ by their estimators $\widehat{\boldsymbol{\theta}}$, $\widehat{x_i}$ and $\widehat{x_i^2}$, $i = 1, \ldots, n$, in the explicit expression of $\boldsymbol{\Omega}$. For the corrected score method in Sects. 2.5 and 3.4, we use the so-called sandwich estimator. With this method, the matrices $\boldsymbol{A}$ and $\boldsymbol{V}$ are replaced by $n^{-1}\sum_{i=1}^n -\partial\boldsymbol{\psi}_i/\partial\boldsymbol{\theta}$ and $n^{-1}\sum_{i=1}^n \boldsymbol{\psi}_i\boldsymbol{\psi}_i'$, respectively, evaluated at the corrected score estimator $\widetilde{\boldsymbol{\theta}}$.

# Appendix B
# R Scripts

In this appendix, we provide R scripts to obtain the results in Chap. 5. The measurements should be in the $n \times (r + 1)$ matrix object called `mobs` with the old method in the first column. The variances of the measurement errors corresponding to the columns of `mobs` should be in the $n \times (r + 1)$ matrix object called `mvar`.

```
# Accurate numerical derivatives
library("numDeriv")

# R functions used in the book
source("library-Regression.R")

# N(0, 1) critical value
sig <- 0.05   # Significance level
zcrit <- qnorm(1 - sig / 2)
```

## B.1 Two Methods (Sect. 5.2)

```
# Ripley & Thompson (1987), without obs. # 22
mobs <- read.table("mobsSection52.txt")
mvar <- read.table("mvarSection52.txt")

n <- nrow(mobs)
nr <- ncol(mobs) - 1

# Chi-square critical value
x2crit <- qchisq(1 - sig, 2 * nr)

# Inference
infCS <- eCS(mobs, mvar)  # corrected score
infML <- eML(mobs, mvar, infCS$theta) # ML

# Scatterplot and fitted models
im <- 1 # New method
xdesc <- expression(paste("Determinations by the old method (", mu,
            "g/l)"))
ydesc <- expression(paste("Determinations by the new method (", mu,
```



H. Bolfarine et al., *Regression Models for the Comparison of Measurement Methods*, SpringerBriefs in Statistics, https://doi.org/10.1007/978-3-030-57935-7

```
            "g/l)"))

rxy <- range(mobs[, 1] - sqrt(mvar[, 1]), mobs[, 1] +
         sqrt(mvar[, 1]), mobs[, im + 1] - sqrt(mvar[, im + 1]),
         mobs[, im + 1] + sqrt(mvar[, im + 1]))
par(mai = c(1.1, 1.1, 0.1,0.1))
plot(mobs[, 1], mobs[, im + 1],  pch = 20, cex.lab = 1.5,
     cex.axis = 1.5, xlim = rxy, ylim = rxy, xlab = xdesc, ylab = ydesc)
segments(mobs[, 1] - sqrt(mvar[, 1]), mobs[, im + 1],
         mobs[, 1] + sqrt(mvar[, 1]), mobs[, im + 1], col = "blue")
segments(mobs[, 1], mobs[, im + 1] - sqrt(mvar[, im + 1]),
         mobs[, 1], mobs[, im + 1] + sqrt(mvar[, im + 1]), col = "blue")
abline(infML$theta[im], infML$theta[im + nr], lty = 1)
abline(infCS$theta[im], infCS$theta[im + nr], lty = 2)

# Pivotal quantity and envelope
envpiv <- exactenv(mobs, mvar, infML$theta, nsim = 100)
par(mai = c(0.95, 0.95, 0.5, 0.1))
with(envpiv, {
  plot(tqn, vrs, xlab = "N(0,1) quantiles",
       ylab = "Residuals and simulated envelope",
       pch = 20, ylim = range(vrs, vrmin, vrmax), cex.lab = 1.4,
       cex.axis = 1.4, col = "blue")
  lines(tqn, vrmin)
  lines(tqn, vrm)
  lines(tqn, vrmax)
})

# Exact test
infexact <- exact(mobs, mvar)

## Confidence regions
npoints <- 500
confCS <- confregion(infCS$covm, infCS$theta, sqrt(x2crit), npoints)
confML <- confregion(infML$covm, infML$theta, sqrt(x2crit), npoints)
rx <- range(0, confCS[, 1], confML[, 1])
ry <- range(1, confCS[, 2], confML[, 2])

npoints <- 100
ra <- seq(rx[1], rx[2], length = npoints)
rb <- seq(ry[1], ry[2], length = npoints)
confZ <- matrix(0, npoints, npoints)
for (i in 1:npoints)
{
  for (j in 1:npoints) {
    confZ[i, j] <- abs(pivot(mobs, mvar, c(ra[i], rb[j])))
  }
}

par(mai = c(0.95, 0.95, 0.5, 0.1))
plot(confML[, 1], confML[, 2], type = "l", xlab = expression(alpha),
     ylab = expression(beta), xlim = rx, ylim = ry, cex.lab = 1.4,
     cex.axis = 1.4)
lines(confCS[, 1], confCS[, 2], lty = 2)
contour(ra, rb, confZ, lty = 3, levels = zcrit, drawlabels = FALSE,
        add = TRUE)
points(0, 1, pch = 20)

# Probability of agreement plot
ohatx <- order(infML$hatx)
im <- 1  # New method
c0 <- 0.54 # Tolerance
mypoa <- poa(mvar[, 1], mvar[, im + 1], c0 = c0,
    infML$hatx, infML$theta[c(im, im + nr)],
    infML$covm[c(im, im + nr), c(im, im + nr)], 0.95)
mypoaH0 <- poa(mvar[, 1], mvar[, im + 1], c0 = c0, infML$hatx, c(0, 1),
    infML$covm[c(im, im + nr), c(im, im + nr)], 0.95) # Under H0
```

```
par(mai = c(0.95, 0.95, 0.5, 0.1))
plot(infML$hatx, mypoa$poa, type = "n", ylim = c(0, 1),
  ylab = "Probability of agreement", cex.lab = 1.4, cex.axis = 1.4,
  xlab = expression(hat(x)))
lines(infML$hatx[ohatx], mypoa$poa[ohatx])
points(infML$hatx, mypoa$poa, pch = 20, cex = 0.8)
points(infML$hatx, mypoaH0$poa, pch = 20, col = "blue")
segments(infML$hatx, mypoa$lower, infML$hatx, mypoa$upper, col = "red")
rug(infML$hatx)

## Local influence
myinf <- localinfluence(mobs, mvar, infML$theta)

# Sample weights
# Wald
rab <- range(abs(myinf$dmaxt0sw), abs(myinf$dmaxzsw))
par(mai = c(1.0, 1.0, 0.1,0.15))
plot(abs(myinf$dmaxt0sw), pch = 1, ylab = expression(paste("|", d[max],
   "|")), cex.lab = 1.4, cex.axis = 1.4, col = "blue", axes = FALSE,
   ylim = rab)
text(19, abs(myinf$dmaxt0sw)[19], "19", cex = 1.2)
text(22, abs(myinf$dmaxt0sw)[22], "22", cex = 1.2)
axis(1, 1:n, cex.axis = 1.4)
axis(2, cex.axis = 1.4)
box()

# Pivotal quantity
par(mai = c(1.0, 1.0, 0.1,0.15))
plot(abs(myinf$dmaxzsw), pch = 1, ylab = expression(paste("|", d[max],
   "|")), cex.lab = 1.4, cex.axis = 1.4, col = "blue", axes = FALSE,
   ylim = rab)
text(13, abs(myinf$dmaxzsw)[13], "13", cex = 1.2)
text(19, abs(myinf$dmaxzsw)[19], "19", cex = 1.2)
text(22, abs(myinf$dmaxzsw)[22], "22", cex = 1.2)
text(23, abs(myinf$dmaxzsw)[23], "23", cex = 1.2)
text(24, abs(myinf$dmaxzsw)[24], "24", cex = 1.2)
axis(1, 1:n, cex.axis = 1.4)
axis(2, cex.axis = 1.4)
box()

## Test statistics after removing one sample a time
Wi <- Zi <- c()
for (i in 1:n) {
  Wi[i] <- eML(mobs[-i,], mvar[-i,], theta = infML$theta,
               verbose = FALSE)$W
  Zi[i] <- abs(exact(mobs[-i,], mvar[-i,], verbose = FALSE)$Z0)
}

# Wald
rab <- range(infML$W, Wi, x2crit)
par(mai = c(1.0, 1.0, 0.1,0.15))
plot(Wi, pch = 1, ylab = expression(W[-i]), cex.lab = 1.4,
    cex.axis = 1.4, col = "blue", axes = FALSE, ylim = rab)
abline(h = x2crit, lty = 2, col = "red")
abline(h = infML$W, lty = 2)
text(n / 2, x2crit, "Critical value", cex = 1.2)
text(n / 2, infML$W, "W", cex = 1.2)
text(7, Wi[7], "7", cex = 1.2)
text(19, Wi[19], "19", cex = 1.2)
text(22, Wi[22], "22", cex = 1.2)
text(23, Wi[23], "23", cex = 1.2)
text(24, Wi[24], "24", cex = 1.2)
axis(1, 1:n, cex.axis = 1.4)
axis(2, cex.axis = 1.4)
box()

# Pivotal quantity
```

```
rab <- range(abs(infexact$Z0) , Zi, zcrit)
par(mai = c(1.0, 1.0, 0.1,0.15))
plot(Zi, pch = 1, ylab = expression(paste("|", Z[0]["."][-i], "|")),
     cex.lab = 1.4, cex.axis = 1.4,  col = "blue", axes = FALSE,
     ylim = rab)
abline(h = zcrit, lty = 2, col = "red")
abline(h = abs(infexact$Z0), lty = 2)
text(n / 2, zcrit, "Critical value", cex = 1.2)
text(n / 2, abs(infexact$Z0), expression(paste("|", Z[0], "|")),
     cex = 1.2)
text(13, Zi[13], "13", cex = 1.2)
text(19, Zi[19], "19", cex = 1.2)
text(22, Zi[22], "22", cex = 1.2)
text(23, Zi[23], "23", cex = 1.2)
text(24, Zi[24], "24", cex = 1.2)
axis(1, 1:n, cex.axis = 1.4)
axis(2, cex.axis = 1.4)
box()

## Old method - additive perturbation
rab <- range(abs(myinf$dmaxt0aom), abs(myinf$dmaxzaom),
             abs(myinf$dmaxt0mom), abs(myinf$dmaxzmom))

# Wald
par(mai = c(1.0, 1.0, 0.1,0.15))
plot(abs(myinf$dmaxt0aom), pch = 1, ylab = expression(paste("|", d[max],
    "|")), cex.lab = 1.4, cex.axis = 1.4, col = "blue", axes = FALSE,
     ylim = rab)
text(7, abs(myinf$dmaxt0aom)[7], "7", cex = 1.2)
text(19, abs(myinf$dmaxt0aom)[19], "19", cex = 1.2)
text(22, abs(myinf$dmaxt0aom)[22], "22", cex = 1.2)
text(23, abs(myinf$dmaxt0aom)[23], "23", cex = 1.2)
text(24, abs(myinf$dmaxt0aom)[24], "24", cex = 1.2)
axis(1, 1:n, cex.axis = 1.4)
axis(2, cex.axis = 1.4)
box()

# Pivotal quantity
par(mai = c(1.0, 1.0, 0.1,0.15))
plot(abs(myinf$dmaxzaom), pch = 1, ylab = expression(paste("|", d[max],
    "|")), cex.lab = 1.4, cex.axis = 1.4, col = "blue", axes = FALSE,
     ylim = rab)
text(7, abs(myinf$dmaxzaom)[7], "7", cex = 1.2)
text(19, abs(myinf$dmaxzaom)[19], "19", cex = 1.2)
text(22, abs(myinf$dmaxzaom)[22], "22", cex = 1.2)
text(24, abs(myinf$dmaxzaom)[24], "24", cex = 1.2)
axis(1, 1:n, cex.axis = 1.4)
axis(2, cex.axis = 1.4)
box()

## Old method - multiplicative perturbation
# Wald
par(mai = c(1.0, 1.0, 0.1,0.15))
plot(abs(myinf$dmaxt0mom), pch = 1, ylab = expression(paste("|", d[max],
    "|")), cex.lab = 1.4, cex.axis = 1.4, col = "blue", axes = FALSE,
     ylim = rab)
text(7, abs(myinf$dmaxt0mom)[7], "7", cex = 1.2)
text(19, abs(myinf$dmaxt0mom)[19], "19", cex = 1.2)
axis(1, 1:n, cex.axis = 1.4)
axis(2, cex.axis = 1.4)
box()

# Pivotal quantity
par(mai = c(1.0, 1.0, 0.1,0.15))
plot(abs(myinf$dmaxzmom), pch = 1, ylab = expression(paste("|", d[max],
    "|")), cex.lab = 1.4, cex.axis = 1.4, col = "blue", axes = FALSE,
     ylim = rab)
```

```
axis(1, 1:n, cex.axis = 1.4)
axis(2, cex.axis = 1.4)
box()

## New method - additive perturbation
rab <- range(abs(myinf$dmaxt0anm), abs(myinf$dmaxzanm),
             abs(myinf$dmaxt0mnm), abs(myinf$dmaxzmnm))
# Wald
par(mai = c(1.0, 1.0, 0.1,0.15))
plot(abs(myinf$dmaxt0anm), pch = 1, ylab = expression(paste("|", d[max],
    "|")), cex.lab = 1.4, cex.axis = 1.4, col = "blue", axes = FALSE,
     ylim = rab)
text(7, abs(myinf$dmaxt0anm)[7], "7", cex = 1.2)
text(19, abs(myinf$dmaxt0anm)[19], "19", cex = 1.2)
text(22, abs(myinf$dmaxt0anm)[22], "22", cex = 1.2)
text(24, abs(myinf$dmaxt0anm)[24], "24", cex = 1.2)
axis(1, 1:n, cex.axis = 1.4)
axis(2, cex.axis = 1.4)
box()

# Pivotal quantity
par(mai = c(1.0, 1.0, 0.1,0.15))
plot(abs(myinf$dmaxzanm), pch = 1, ylab = expression(paste("|", d[max],
    "|")), cex.lab = 1.4, cex.axis = 1.4, col = "blue", axes = FALSE,
     ylim = rab)
text(7, abs(myinf$dmaxzanm)[7], "7", cex = 1.2)
text(19, abs(myinf$dmaxzanm)[19], "19", cex = 1.2)
text(22, abs(myinf$dmaxzanm)[22], "22", cex = 1.2)
text(24, abs(myinf$dmaxzanm)[24], "24", cex = 1.2)
axis(1, 1:n, cex.axis = 1.4)
axis(2, cex.axis = 1.4)
box()

## New method - multiplicative perturbation
# Wald
par(mai = c(1.0, 1.0, 0.1,0.15))
plot(abs(myinf$dmaxt0mnm), pch = 1, ylab = expression(paste("|", d[max],
    "|")), cex.lab = 1.4, cex.axis = 1.4, col = "blue", axes = FALSE,
     ylim = rab)
text(7, abs(myinf$dmaxt0mnm)[7], "7", cex = 1.2)
text(19, abs(myinf$dmaxt0mnm)[19], "19", cex = 1.2)
text(22, abs(myinf$dmaxt0mnm)[22], "22", cex = 1.2)
text(24, abs(myinf$dmaxt0mnm)[24], "24", cex = 1.2)
axis(1, 1:n, cex.axis = 1.4)
axis(2, cex.axis = 1.4)
box()

# Pivotal quantity
par(mai = c(1.0, 1.0, 0.1,0.15))
plot(abs(myinf$dmaxzmnm), pch = 1, ylab = expression(paste("|", d[max],
    "|")), cex.lab = 1.4, cex.axis = 1.4, col = "blue", axes = FALSE,
     ylim = rab)
axis(1, 1:n, cex.axis = 1.4)
axis(2, cex.axis = 1.4)
box()

# Old method, additive perturbation
# Direction d_max(22, 24) with a = -0.3
dmaxt0 <- myinf$dmaxt0aom
dmaxt0[-c(22, 24)] <- 0
dmaxt0 <- dmaxt0 / sqrt(sum(dmaxt0^2))
a <- -0.3
tmobs <- cbind(mobs[, 1] + a * dmaxt0, mobs[, -1])
cat("\n X(22, 24) =", tmobs[c(22, 24), 1], "\n")
Wa <- eML(tmobs, mvar, theta = infML$theta, verbose = FALSE)$W
cat("\n a =", a, ": W =", Wa, ", d.f.=", nr + 1, " (p = ",
    pchisq(Wa, df = nr + 1, lower.tail = FALSE), ") \n")
```

## B.2 Three Methods (Sect. 5.3)

```
# Lao et al. (1996, Table 5)
mobs <- read.table("mobsSection53.txt")
mvar <- read.table("mvarSection53.txt")

n <- nrow(mobs)
nr <- ncol(mobs) - 1

# Chi-square critical value
x2crit <- qchisq(1 - sig, 2 * nr)

# Inference
infCS <- eCS(mobs, mvar)  # corrected score
infML <- eML(mobs, mvar, infCS$theta) # ML

# Scatterplots and fitted models
xdesc <- "Recoveries by the MAP method (%)"

im <- 1 # New method
ydesc <- "Recoveries by the Soxhlet extraction method (%)"
rxy <- range(mobs[, 1] - sqrt(mvar[, 1]), mobs[, 1] +
         sqrt(mvar[, 1]), mobs[, im + 1] - sqrt(mvar[, im + 1]),
         mobs[, im + 1] + sqrt(mvar[, im + 1]))
par(mai = c(1.1, 1.1, 0.1,0.1))
plot(mobs[, 1], mobs[, im + 1],  pch = 20, cex.lab = 1.5,
     cex.axis = 1.5, xlim = rxy, ylim = rxy, xlab = xdesc, ylab = ydesc)
segments(mobs[, 1] - sqrt(mvar[, 1]), mobs[, im + 1],
         mobs[, 1] + sqrt(mvar[, 1]), mobs[, im + 1], col = "blue")
segments(mobs[, 1], mobs[, im + 1] - sqrt(mvar[, im + 1]),
         mobs[, 1], mobs[, im + 1] + sqrt(mvar[, im + 1]), col = "blue")
abline(infML$theta[im], infML$theta[im + nr], lty = 1)

im <- 2 # New method
ydesc <- "Recoveries by the mechanical extraction method (%)"
rxy <- range(mobs[, 1] - sqrt(mvar[, 1]), mobs[, 1] +
         sqrt(mvar[, 1]), mobs[, im + 1] - sqrt(mvar[, im + 1]),
         mobs[, im + 1] + sqrt(mvar[, im + 1]))
par(mai = c(1.1, 1.1, 0.1,0.1))
plot(mobs[, 1], mobs[, im + 1],  pch = 20, cex.lab = 1.5,
     cex.axis = 1.5, xlim = rxy, ylim = rxy, xlab = xdesc, ylab = ydesc)
segments(mobs[, 1] - sqrt(mvar[, 1]), mobs[, im + 1],
         mobs[, 1] + sqrt(mvar[, 1]), mobs[, im + 1], col = "blue")
segments(mobs[, 1], mobs[, im + 1] - sqrt(mvar[, im + 1]),
         mobs[, 1], mobs[, im + 1] + sqrt(mvar[, im + 1]), col = "blue")
abline(infML$theta[im], infML$theta[im + nr], lty = 1)

# Pivotal quantity and envelope
envpiv <- exactenv(mobs, mvar, infML$theta, nsim = 100)
par(mai = c(0.95, 0.95, 0.5, 0.1))
with(envpiv, {
  plot(tqn, vrs, xlab = "N(0,1) quantiles",
       ylab = "Residuals and simulated envelope",
       pch = 20, ylim = range(vrs, vrmin, vrmax), cex.lab = 1.4,
       cex.axis = 1.4, col = "blue")
  lines(tqn, vrmin)
  lines(tqn, vrm)
  lines(tqn, vrmax)
})

# Probability of agreement plots
ohatx <- order(infML$hatx)

im <- 1  # New method
c0 <- 4.2
mypoa <- poa(mvar[, 1], mvar[, im + 1], c0 = c0,
```

```
    infML$hatx, infML$theta[c(im, im + nr)],
    infML$covm[c(im, im + nr), c(im, im + nr)], 0.95)
mypoaH0 <- poa(mvar[, 1], mvar[, im + 1], c0 = c0, infML$hatx, c(0, 1),
    infML$covm[c(im, im + nr), c(im, im + nr)], 0.95) # Under H0
par(mai = c(0.95, 0.95, 0.5, 0.1))
plot(infML$hatx, mypoa$poa, type = "n", ylim = c(0, 1),
  ylab = "Probability of agreement", cex.lab = 1.4, cex.axis = 1.4,
  xlab = expression(hat(x)))
lines(infML$hatx[ohatx], mypoa$poa[ohatx])
points(infML$hatx, mypoa$poa, pch = 20, cex = 0.8)
points(infML$hatx, mypoaH0$poa, pch = 20, col = "blue")
segments(infML$hatx, mypoa$lower, infML$hatx, mypoa$upper, col = "red")
rug(infML$hatx)

im <- 2  # New method
c0 <- 15.95
mypoa <- poa(mvar[, 1], mvar[, im + 1], c0 = c0,
    infML$hatx, infML$theta[c(im, im + nr)],
    infML$covm[c(im, im + nr), c(im, im + nr)], 0.95)
mypoaH0 <- poa(mvar[, 1], mvar[, im + 1], c0 = c0, infML$hatx, c(0, 1),
    infML$covm[c(im, im + nr), c(im, im + nr)], 0.95) # Under H0
par(mai = c(0.95, 0.95, 0.5, 0.1))
plot(infML$hatx, mypoa$poa, type = "n", ylim = c(0, 1),
  ylab = "Probability of agreement", cex.lab = 1.4, cex.axis = 1.4,
  xlab = expression(hat(x)))
lines(infML$hatx[ohatx], mypoa$poa[ohatx])
points(infML$hatx, mypoa$poa, pch = 20, cex = 0.8)
points(infML$hatx, mypoaH0$poa, pch = 20, col = "blue")
segments(infML$hatx, mypoa$lower, infML$hatx, mypoa$upper, col = "red")
rug(infML$hatx)

## Local influence
myinf <- localinfluence(mobs, mvar, infML$theta)

# Sample weights
par(mai = c(1.0, 1.0, 0.1,0.15))
plot(abs(myinf$dmaxt0sw), pch = 1, ylab = expression(paste("|", d[max],
  "|")), cex.lab = 1.4, cex.axis = 1.4, col = "blue", axes = FALSE)
text(2, abs(myinf$dmaxt0sw)[2], "2", cex = 1.2)
text(11, abs(myinf$dmaxt0sw)[11], "11", cex = 1.2)
text(21, abs(myinf$dmaxt0sw)[21], "21", cex = 1.2)
axis(1, 1:n, cex.axis = 1.4)
axis(2, cex.axis = 1.4)
box()

# Wald statistic after removing one sample a time
Wi <- c()
for (i in 1:n) {
  Wi[i] <- eML(mobs[-i,], mvar[-i,], theta = infML$theta,
               verbose = FALSE)$W
}
rab <- range(infML$W, Wi, x2crit)
par(mai = c(1.0, 1.0, 0.1,0.15))
plot(Wi, pch = 1, ylab = expression(W[-i]), cex.lab = 1.4,
    cex.axis = 1.4, col = "blue", axes = FALSE, ylim = rab)
abline(h = x2crit, lty = 2, col = "red")
abline(h = infML$W, lty = 2)
text(n / 2, x2crit, "Critical value", cex = 1.2)
text(n / 2, infML$W, "W", cex = 1.2)
text(2, Wi[2], "2", cex = 1.2)
text(11, Wi[11], "11", cex = 1.2)
text(21, Wi[21], "21", cex = 1.2)
axis(1, 1:n, cex.axis = 1.4)
axis(2, cex.axis = 1.4)
box()

# Old method - additive perturbation
```

```
rab <- range(abs(myinf$dmaxt0aom), abs(myinf$dmaxt0mom))
par(mai = c(1.0, 1.0, 0.1,0.15))
plot(abs(myinf$dmaxt0aom), pch = 1, ylab = expression(paste("|", d[max],
    "|")), cex.lab = 1.4, cex.axis = 1.4, col = "blue", axes = FALSE,
     ylim = rab)
text(13, abs(myinf$dmaxt0aom)[13], "13", cex = 1.2)
axis(1, 1:n, cex.axis = 1.4)
axis(2, cex.axis = 1.4)
box()

# Old method - multiplicative perturbation
par(mai = c(1.0, 1.0, 0.1,0.15))
plot(abs(myinf$dmaxt0mom), pch = 1, ylab = expression(paste("|", d[max],
    "|")), cex.lab = 1.4, cex.axis = 1.4, col = "blue", axes = FALSE,
     ylim = rab)
text(13, abs(myinf$dmaxt0mom)[13], "13", cex = 1.2)
axis(1, 1:n, cex.axis = 1.4)
axis(2, cex.axis = 1.4)
box()

# New methods - additive and multiplicative perturbations
rab <- range(abs(myinf$dmaxt0anm), abs(myinf$dmaxt0mnm))
im <- 1 # New method
par(mai = c(1.0, 1.0, 0.1,0.15))
plot(abs(myinf$dmaxt0anm[, im]), pch = 1, ylab = expression(paste("|",
     d[max], "|")), cex.lab = 1.4, cex.axis = 1.4, col = "blue",
     axes = FALSE, ylim = rab)
text(11, abs(myinf$dmaxt0anm)[11, im], "11", cex = 1.2)
text(21, abs(myinf$dmaxt0anm)[21, im], "21", cex = 1.2)
axis(1, 1:n, cex.axis = 1.4)
axis(2, cex.axis = 1.4)
box()

par(mai = c(1.0, 1.0, 0.1,0.15))
plot(abs(myinf$dmaxt0mnm[, im]), pch = 1, ylab = expression(paste("|",
     d[max], "|")), cex.lab = 1.4, cex.axis = 1.4, col = "blue",
     axes = FALSE, ylim = rab)
text(11, abs(myinf$dmaxt0anm)[11, im], "11", cex = 1.2)
text(21, abs(myinf$dmaxt0anm)[21, im], "21", cex = 1.2)
axis(1, 1:n, cex.axis = 1.4)
axis(2, cex.axis = 1.4)
box()

im <- 2 # New method
par(mai = c(1.0, 1.0, 0.1,0.15))
plot(abs(myinf$dmaxt0anm[, im]), pch = 1, ylab = expression(paste("|",
     d[max], "|")), cex.lab = 1.4, cex.axis = 1.4, col = "blue",
     axes = FALSE, ylim = rab)
text(21, abs(myinf$dmaxt0anm)[21, im], "21", cex = 1.2)
axis(1, 1:n, cex.axis = 1.4)
axis(2, cex.axis = 1.4)
box()

par(mai = c(1.0, 1.0, 0.1,0.15))
plot(abs(myinf$dmaxt0mnm[, im]), pch = 1, ylab = expression(paste("|",
     d[max], "|")), cex.lab = 1.4, cex.axis = 1.4, col = "blue",
     axes = FALSE, ylim = rab)
text(21, abs(myinf$dmaxt0anm)[21, im], "21", cex = 1.2)
axis(1, 1:n, cex.axis = 1.4)
axis(2, cex.axis = 1.4)
box()
```

# References


Altman DG, Bland JM (1983) Measurement in medicine: the analysis of method comparison studies. Statistician 32:307–317

Atkinson AC (1985) Plots, transformations, and regression. Clarendon, Oxford

Barnett V (1969) Simultaneous pairwise linear structural relationships. Biometrics 25:129–142

Bland JM, Altman DG (1986) Statistical methods for assessing agreement between two methods of clinical measurement. Lancet 327:307–310

Cadigan NG, Farrell PJ (2002) Generalized local influence with applications to fish stock cohort analysis. Appl Statist 51:469–483

Carroll RJ, Ruppert D, Stefanski LA, Crainiceanu CM (2006) Measurement error in nonlinear models: a modern perspective, 2nd edn. Chapman & Hall/CRC, Boca Raton

Carstensen B (2010) Comparing clinical measurement methods. Wiley, Chichester

Cheng CL, Riu J (2006) On estimating linear relationships when both variables are subject to heteroscedastic measurement errors. Technometrics 48(4):511–519

Cheng CL, Van Ness JW (1999) Statistical regression with measurement error. Arnold, London

Choudhary PK, Nagaraja HN (2017) Measuring agreement: models, methods, and applications. Wiley, Hoboken

Cook RD (1986) Assessment of local influence (with discussion). J R Stat Soc B 48:133–169

Cook RD (1987) Influence assessment. J Appl Statist 14:117–131

de Castilho MV (2004) A comparison of statistical techniques for detecting analytical bias in geoanalysis. Geostand Geoanal Res 28:277–290

de Castro M, Galea M (2010) Robust inference in an heteroscedastic measurement error model. J Korean Statist Soc 39:439–447

de Castro M, Galea-Rojas M, Bolfarine H, de Castilho MV (2004) Detection of analytical bias when comparing two or more measuring devices. J Chemom 18:431–440

de Castro M, Bolfarine H, Galea-Rojas M, de Castilho MV (2005) An exact test for analytical bias detection. Analytica Chimica Acta 538:375–381

de Castro M, de Castilho MV, Bolfarine H (2006a) Consistent estimation and testing in comparing analytical bias models. Environmetrics 17:167–182

de Castro M, Galea-Rojas M, Bolfarine H, de Castilho MV (2006b) Local influence in regression models for the detection of analytical bias. Chemom Intell Lab Syst 83:139–147

de Castro M, Galea-Rojas M, Bolfarine H (2007) Local influence assessment in heteroscedastic measurement error models. Comput Stat Data Anal 52:1132–1142

del Río FJ, Riu J, Rius FX (2001) Prediction intervals in linear regression taking into account errors on both axes. J Chemother 15:773–788





H. Bolfarine et al., *Regression Models for the Comparison of Measurement Methods*, SpringerBriefs in Statistics, https://doi.org/10.1007/978-3-030-57935-7

Dolby GR, Cormack RM, Sinclair DF (1987) On fitting bivariate functional relationships to unpaired and unequally replicated data. Biometrika 74:393–399
Dunn G (2004) Design and analysis of reliability: the statistical evaluation of measurement errors, 2nd edn. Arnold, New York
Ermer J, Miller JHM (2005) Method validation in pharmaceutical analysis: a guide to best practice. Wiley, Weinheim
Escobar LA, Meeker WQ (1992) Assessing influence in regression analysis with censored data. Biometrics 48:507–528
Fang KT, Kotz S, Ng KW (1990) Symmetric multivariate and related distributions. Chapman and Hall, London
Fuller WA (1987) Measurement error models. Wiley, New York
Galea M, de Castro M (2017) Robust inference in a linear functional model with replications using the $t$ distribution. J Multivar Anal 160:134–145
Galea-Rojas M, de Castilho MV, Bolfarine H, de Castro M (2003) Detection of analytical bias. Analyst 128:1073–1081
Gilbert P, Varadhan R (2016) numDeriv: accurate numerical derivatives. r package version 2016.8-1. http://CRAN.R-project.org/package=numDeriv
Gimenez P, Bolfarine H (1997) Corrected score functions in classical error-in-variables and incidental parameter models. Aust J. Statist 39:325–344
Kimura DK (1992) Functional comparative calibration using an EM algorithm. Biometrics 48:1263–1271
Kukush A, Van Huffel S (2004) Consistency of elementwise-weighted total least squares estimator in a multivariate errors-in-variables model AX=B. Metrika 59:75–97
Kukush A, Markovsky I, Huffel SV (2002) On consistent estimators in linear and bilinear multivariate errors-in-variables models. In: Van Huffel S, Lemmerling P (eds) Total least squares and errors-in-variables modeling: analysis, algorithms and applications. Kluwer Academic Publishers, Dordrecht, pp 157–166
Lao RC, Shu YY, Holmes J, Chiu C (1996) Environmental sample cleaning and extraction procedures by microwave-assisted process (MAP) technology. Microchem J 53:99–108
Lee SY, Wang S (1996) Sensitivity analysis of structural equation models. Psychometrika 61:93–108
Lehmann EL (1998) Elements of large-sample theory. Springer, New York
Lin L, Hedayat A, Wu W (2012) Statistical tools for measuring agreement. Springer, New York
Lisý JM, Cholvadová A, Kutej J (1990) Multiple straight-line least-squares analysis with uncertainties in all variables. Comput Chem 14:189–192
Mak TK (1983) On Sprent's generalized least-squares estimator. J R Stat Soc B 45:380–383
Markovsky I, Rastello ML, Premoli A, Kukush A, Van Huffel S (2006) The element-wise weighted total least squares problem. Comput Statist Data Anal 50:181–209
Martínez A, del Río FJ, Riu J, Rius FX (1999) Detecting proportional and constant bias in method comparison studies by using linear regression with errors in both axes. Chemom Intell Lab Syst 49:179–193
Meier R, Zünd E (2000) Statistical methods in analytical chemistry, 2nd edn. Wiley, New York
Nakamura T (1990) Corrected score function of errors-in-variables models: methodology and applications to generalized linear models. Biometrika 77:127–137
R Core Team (2019) R: a language and environment for statistical computing. R Foundation for Statistical Computing, Vienna
Ripley BD, Thompson M (1987) Regression techniques for the detection of analytical bias. Analyst 112:377–383
Riu J, Rius FX (1996) Assessing the accuracy of analytical methods using linear regression with errors in both axes. Anal Chem 68:1851–1857
Sen PK, Singer JM (1993) Large sample methods in statistics: an introduction with applications. Chapman & Hall, New York
Shoukri MM (2011) Measures of interobserver agreement and reliability, 2nd edn. CRC Press, Boca Raton

Smyth GK (1996) Partitioned algorithms for maximum likelihood and other non-linear estimation. Statist Comput 6:201–216
Sprent P (1966) A generalized least-squares approach to linear functional relationships. J R Statist Soc B 28:278–297
St Laurent RT (1998) Evaluating agreement with a gold standard in method comparison studies. Biometrics 54:537–545
Stefanski LA (1989) Unbiased estimation of a nonlinear function of a normal mean with application to measurement error models. Commun Statist Theory Methods 18:4335–4358
Stevens NT, Steiner SH, MacKay RJ (2017) Assessing agreement between two measurement systems: an alternative to the limits of agreement approach. Statist Methods Med Res 26:2487–2504
Thisted RA (1988) Elements of statistical computing. Chapman & Hall, New York
Verbeke G, Molenberghs G (2000) Linear mixed models for longitudinal data. Springer, New York
Walter SD (1997) Variation in baseline risk as an explanation of heterogeneity in meta-analysis. Statist Med 16:2883–2900
Woodhouse R (2003) Statistical regression line-fitting in the oil and gas industry. Penn Well, Tulsa
Wu X, Luo Z (1993) Second-order approach to local influence. J R Statist Soc B 55:929–936

# Index





H. Bolfarine et al., *Regression Models for the Comparison of Measurement Methods*, SpringerBriefs in Statistics, https://doi.org/10.1007/978-3-030-57935-7